WHAT IS
VEGANISM
FOR?

The status quo is broken. Humanity today faces multiple interconnected challenges, some of which could prove existential. If we believe the world could be different, if we want it to be *better*, examining the purpose of what we do – and what is done in our name – is more pressing than ever.

The What Is It For? series examines the purpose of the most important aspects of our contemporary world, from religion and free speech to animal rights and the Olympics. It illuminates what these things are by looking closely at what they do.

The series offers fresh thinking on current debates that gets beyond the overheated polemics and easy polarizations. Across the series, leading experts explore new ways forward, enabling readers to engage with the possibility of real change.

Series editor: George Miller

Visit **bristoluniversitypress.co.uk/what-is-it-for** to find out more about the series.

CATHERINE OLIVER is a geographer and Lecturer in the Sociology of Climate Change at Lancaster University. She writes about veganism, chickens and other animals, having previously worked at the University of Cambridge researching urban chickens in London. She is the author of *Veganism, Archives, and Animals* (Routledge, 2021) and lives on the edge of Morecambe Bay with her guinea pigs.

WHAT IS
VEGANISM
FOR?

CATHERINE OLIVER

First published in Great Britain in 2024 by

Bristol University Press
University of Bristol
1–9 Old Park Hill
Bristol
BS2 8BB
UK
t: +44 (0)117 374 6645
e: bup-info@bristol.ac.uk

Details of international sales and distribution partners are available at
bristoluniversitypress.co.uk

British Library Cataloguing in Publication Data
A catalogue record for this book is available from the British Library

ISBN 978-1-5292-3432-9 paperback
ISBN 978-1-5292-3433-6 ePub
ISBN 978-1-5292-3434-3 ePdf

Cover design: Tom Appshaw

CONTENTS

LIST OF FIGURES AND BOXES

Figures

Boxes

PROLOGUE

Ten years ago, I watched a documentary called *Earthlings*. At the time, *Earthlings* had a notoriety among vegans, sometimes referred to as 'the vegan-maker' because of the violent footage it contains. The film is narrated by vegan actor Joaquin Phoenix and shows in horrifying detail the exploitation, lives and deaths of animals in puppy mills, factory farms, research labs, entertainment and the fashion industry. I went vegan the next day. The next few years were a steep learning curve for me in removing animal products from my diet, wardrobe, beauty and healthcare routines, and all sorts of other hidden places.

At the same time, my eyes were being opened to the impacts of animal agriculture on the environment, from soil degradation to toxic pollution. I also began researching the health impacts of eating animals and the benefits of eating plants. It was sometimes difficult to decipher fact from fad, as veganism became seen less as an extremist position and instead part of the 'wellness' trends of the 2010s, and slowly made its way into environmentalist discourses. But with growing visibility came new difficulties for veganism, still dismissed in popular culture as the territory of wealthy White women or an unrealistic endeavour.

Over the last decade, the place of veganism in society, culture and politics has transformed. With changing ethical, environmental and health challenges, what we eat, wear and consume is being reconsidered on a larger scale than ever before as we face new planetary pressures. Veganism is becoming increasingly important in these conversations, though it is not a magic solution to the world's issues, and nor is it yet fully understood as *more than a diet*.[1] Across the world, many people are looking to change how they live, because they believe our current way of life is unsustainable and unethical for humans, animals and the planet. Some of them are turning to veganism. I have witnessed mounting evidence for the importance of veganism in thinking about not just the future of food, but the future of humanity, ethics and the environment.

But while veganism is certainly taken more seriously than ever before, it often still evokes confusion, dismissal and even anger. This is at least in part because food isn't simply about what we eat: it is a social, cultural and community-building activity. Rejecting the consumption of animals can lead to our families, friends and communities feeling rejected or criticized. What we choose to eat represents our values as individuals, groups and societies: we are *how* we eat, and the presence of the vegan at the table changes the atmosphere,[2] even if it isn't always obvious why.

This book is about veganism. In it, I explore veganism's claims to offer solutions to some of the world's greatest problems through the distinctive set of ethical, political and environmental commitments that define it. I look at veganism's history and development; consider its global spread; and ultimately ask: *what is veganism for?* This book invites readers to think differently about the role of veganism in our pasts but also, crucially, in our futures – as humans, animals and earthly beings.

I would like to thank George Miller and the team at Bristol University Press, as well as my colleagues across the world, for their generous support and guidance in my work. I'd also like to thank my family, friends, partner and three beloved guinea pigs, Parsley, Sage and Cosmo, without whom the future would be hopeless.

1

INTRODUCTION

In 1944, English animal rights advocates Donald and Dorothy Watson founded the Vegan Society. Removing all animal products and byproducts from their diets, they coined the term 'veganism' to describe their new diet and began promoting the health and spiritual benefits they experienced, while also advocating for the environmental impacts of this abstention. The Watsons connected the vegan diet to issues such as soil quality, national self-sufficiency and an improved moral character. This British organization of people who had given up animal products is often pinpointed as the beginning of veganism as we know it today: an ethical movement organized around eliminating human consumption of animals (see Figure 1.1). But this perception of veganism as a movement originating in Britain is not straightforward: while the term was coined there, there have been many ways of life that refuse the use of animals for ethical or religious reasons (such as Buddhism and Jainism) throughout history.

Today, veganism is sometimes perceived as a Western export across the world, but in Africa, for example, the prevalence and history of plant-based eating is deeply embedded in traditional ways of life. Chef and author of the book *Vegan Africa*, Marie Kacouchia, has said that in the Ivory Coast, there is no one local calling themselves vegan, but many people who committed to a plant-based diet did so because it was traditional, but 'they will not label themselves vegan, because ... it's not something that is in their representation'.[1] Investment ventures like the Vegan Africa Fund want to capitalize not just on the rich plant-based heritage of Africa, but also the availability of arable land on the African continent, aiming to attract investment that might one day enable Africa to feed the entire planet on a vegan diet. For South Africa-based chef Nicola Kagoro, author of *African Vegan on a Budget*, plant-based eating is less a rise than a return: 'more Africans are going back to being vegan, not becoming vegan',[2] challenging ideas of veganism as originating in Britain in 1944.

The growth of veganism

The ethical, political and spiritual refusal to consume animals *and* the embracing of plants as nutritious have long histories, but the last decade has seen veganism transform. What was once a fringe practice is now a major presence in both cultural discourse and consumption. This transformation isn't geographically limited. Statistics about dietary consumption suggest between 1 and 2 per cent of the United Kingdom are vegan (between 670,000 and 1,340,000 people);[3] across Europe, a 2023 study reported that 2.6 million people were vegan, which has doubled since 2019;[4] 13 per cent of consumers in Asia reported being vegan in 2020 according to one study;[5] and in China in 2022, 5 per cent of the population reported that they were vegan, representing around 70 million people.[6] In the United States in 2023, estimates of vegans are around 1 per cent or about three million people.[7] In India, an estimated five million people are vegans, mostly made up from the country's Hindu populations (of which there are around 500 million vegetarians).[8] In 2023, it is conservatively estimated that 1.1 per cent of the world's total population was vegan, and the growth of veganism shown in continental-level surveys suggests that global numbers have likely doubled in the last five years.

These shifts to veganism are within a wider context of a growing scientific and public critique of what we eat, and how it affects health and sustainability. Shifting attitudes to food indicate cultures are becoming more receptive to veganism as a diet: nearly 60 per cent of people in China are eating less meat and dairy; a survey found 90 per cent of people in South America would be interested in a 'more plant-based diet'; and two in three Americans have reduced their meat intake.[9] Vegan product launches have also been the fastest-growing market across the world – indicating the economic potential of plant-based products.

While these statistics evidence the growing power and appeal of veganism, they are only part of the story. It is also important to understand *why* people are becoming vegan, as opposed to vegetarian or reducing their meat intake. To do this, social and market research has surveyed and interviewed people about why they became vegan. A 2019 survey of 13,000 people from across the world found 68 per cent of vegans reporting animals as their motivation, with 17 per cent citing health and about a tenth the environment.[10] A 2020 study in Hong Kong found 40 per cent of people became vegan for reasons related to animals, and almost a quarter of people for both the environment and health.[11] An earlier study from 2014 in the United States looked at lapsed and current vegans, finding that of respondents who 'quit' veganism, 68 per cent were only motivated by their health: of current vegans, motivations were mixed and multiple; 58 per cent were motivated by animal protection, 59 per cent by the environment, and 63 per cent by disgust about meat, and 69 per cent were also interested in their health.[12] Furthermore, animal advocacy researchers at Faunalytics have also reported that transitions to veganism usually happen gradually, with only one in five becoming vegan overnight. Campaigns like Veganuary have successfully capitalized on the growing interest in veganism, encouraging people to pledge to try veganism for January each year and supporting them with online communities, recipes and information about animal suffering. When Veganuary began in 2014, 3,300 people signed up. In 2023, over 700,000 people took part, from almost every country in the world. From a survey of participants in 2022, 36 per cent of the 639,000 participants planned to stay vegan after January.[13]

These numbers indicate a significant growth of veganism across the world. In addition to those who are already vegan, many more commit to transitioning each year, suggesting continued growth into the future. This surge of veganism is contextualized within wider social changes towards more sustainable and healthy diets, which has led to a surge in the plant-based food market, which in turn is often used to illustrate the growth of veganism. However, it is important to emphasize that statistics are always biased and can't tell the full story of the transformation of veganism. Veganism is difficult to survey accurately for two reasons: surveys are usually undertaken or commissioned by institutions with particular agendas (whether pro- or anti-vegan) and so the sample populations are always going to be somewhat skewed. In addition to this, veganism varies across cultures and societies, and how people label themselves may not always match up with the definitions of veganism used in studies. For example, some may use 'veganism' to refer only to their eating habits, while others understand it as entailing the elimination of all animal products, as well as ethical and political shifts. Even accounting for this, there is clearly a consensus that veganism is growing, and with that growth has come significant debate over its place in society and why it matters.

A brief history of veganism

Throughout its history, veganism (at least for some) has been more than just a diet: it is an ethical and political belief system and practice. Becoming vegan isn't just a dietary change; it is also often a commitment to reducing suffering in all its forms. Especially throughout its modern history in the 20th and 21st centuries (the main focus of this book), it has been widely understood and practised as an all-encompassing belief and practice. The growth of veganism over the past decade has been critiqued by longer-standing vegans and activists who view the marketization of veganism through new products as corporate greenwashing – presenting veganism as something that can be bought and consumed, rather than being rooted in justice. As veganism has become more popular, it has received increasing attention and, while its recent surge is difficult to accurately survey, it has centred primarily around food and consumption, not least as consuming plants has become a way of eating sustainably and healthily (see Chapters 3 and 4, respectively). Environmental concerns and a growing interest in healthy eating have been at least as important to veganism's recent growth as animal ethics, and this coalescing of people, planet and animals is used by vegan advocates to build a case for veganism's transformative power. Yet, while veganism has grown, it is within a broader global context of animals being consumed at a greater rate than ever before, and in more intensive ways. Veganism is not *just* about food, but this book, like many other debates around veganism, often turns to food as the site for the strongest case for veganism, both because of the undeniable suffering of animals in agricultural systems, and the potential impact for the planet and human health of a vegan transition.

Over the last century, pressure has mounted on the agricultural sector to produce food ever more cheaply and quickly, due to both a growing human population and the potential for profits of multinational food corporations, with disastrous consequences for life on Earth. Between 1961 and 2014, the consumption of cows more than doubled from 28 million tonnes to 68 million tonnes per year.[14] In 2020, this equated to 293.2 million cows being slaughtered, a number which is significantly lower than the slaughter of sheep, pigs or chickens. But it isn't just the numbers of cows which has increased; through selective breeding and intensification, cows today are much bigger than their historical counterparts/ancestors. Since 1975, cows farmed for beef have increased in weight by approximately 9 pounds (4kg) per year, totalling a whopping 400-pound (181kg) increase by 2016.[15] This allows farmers to produce more meat at lower cost, but increases health risks for cows. It has also led to significant global impacts as the rising number of cows – whose burps and farts emit high levels of the greenhouse gas methane – is warming the climate. Despite stark warnings over the contributions of cow farming to increasing global temperatures, global consumption – and emissions – are set to rise to unprecedented levels by 2050 if no interventions are made.[16]

Historians have traced the overwhelming growth in animal product consumption back to the nutrition transition beginning in mid-19th-century industrial

Britain that focused on increased overseas production and the importation of meat, sugar and wheat. The taste for animal meat led to land across the world being turned over to its production and laid the foundations for today's global industrial agricultural system. Though we know today that our food systems are unsustainable, meat has long persisted as a symbol of wealth and power, and it occupies a politically significant place in society. Increasing meat consumption has seen animal food production become the largest user of land resources, taking up 80 per cent of global agricultural land, which itself totals a staggering half of the world's habitable land. Of course, a growing human population demands more food, but there are limited resources to meet that demand. Currently, in animal-based industrial food systems, these precious resources are not being used effectively: despite animal agriculture's high land use, it only accounts for 17 per cent of global calorific intake.[17] Rearing animals to produce meat is not just ethically questionable, it is an inefficient way to feed a growing population on a resource-depleted planet.

Following the industrialization of food systems, there have been growing concerns from various sectors of society over what we are eating, and how. Flexitarianism, vegetarianism and locavorism have also received significant interest and uptake in recent years as part of debates around sustainable food futures, but the reasons behind them, whether they work and what it might mean for a diet to work for humans, animals, the planet and the economy remains complicated. Nonetheless, veganism is receiving growing mainstream attention as a potential saviour for environmental, ethical and health crises – while its wider commitments beyond food as an ethical and political *way of life* have been ignored. People, animals and the Earth are under threat from environmental degradation and the climate crisis. The ongoing economic crisis at the time of writing has shown how shocks to food production are wrapped up in political and environmental crises. While food has never been more widely available, *access* to food is still deeply unequal and political.

Figure 1.1: Cover of *The Vegan*, summer 1979

Following its formation in 1944, the Vegan Society began publishing a quarterly newsletter edited by Donald Watson under the title *The Vegan News*. It was renamed *The Vegan* in spring 1946 and is still released quarterly today as a magazine, sharing stories, tips, advocacy and recipes.

Multiple veganisms

Veganism doesn't have a single history, and nor does it have a single definition. Many adhere to the definition pioneered by the Vegan Society:[18]

> a philosophy and way of living which seeks to exclude—as far as is possible and practicable—all forms of exploitation of, and cruelty to, animals for food, clothing, or any other purpose; and by extension, promotes the development and use of animal-free alternatives for the benefit of animals, humans and the environment. In dietary terms it denotes the practice of dispensing with all products derived wholly or partly from animals.

However, as veganism has become more deeply entangled with health and environmentalist agendas, its boundaries have shifted. The tripartite focus of veganism that structures this book, for example, has emerged only in veganism's recent history.

Contemporary veganism's growing concern with human health and the environment, and less with social and political justice, has been interpreted by some as a veganism that has lost its radical intent. Yet, as this book will show, veganism remains committed to offering solutions to people questioning contemporary environmental, health and ethical crises around our consumption and exploitation of animals. This can, however, come with difficulties of clashing values and hopes for the future of veganism (see Box 1.1).

Box 1.1: Defining veganism

Vegan scholars Jan Dutkiewicz and Jonathan Dickstein think definitions of veganism have become too complicated. They argue that there should be a shift in the way veganism is collectively defined to focus not on beliefs, but rather on the practice of abstaining from animal-derived products. Under this definition, a vegan could be someone who refuses animal products for any reason and not be connected with any ethical and political beliefs. This means that in some practices of veganism, concern for animals may be absent altogether.

However, this has provoked resistance from those who identify with veganism as a liberation and anti-oppression movement, such as Richard J. White, who argues that veganism's definition should reflect that it *is* a belief, not just a practice. In the United Kingdom, veganism is protected by equality law, but only if it is a philosophical belief (akin to a religious belief), following a landmark case in 2019. Equality law protects people in the workplace from harassment and discrimination so, for example, someone cannot be treated differently or fired because they are vegan. This protection in law *only* applies if someone is vegan for ethical reasons and is important in ensuring vegans are treated equally in society.

Debates over the definition of veganism are not new. Today, many vegans across the world subscribe to the Vegan Society's definition (see page 10), but even this has changed and refined over time. The first definition of veganism was proposed by Leslie J. Cross in 1949, five years after the founding of the Society, and it has continued to be updated since, with today's definition being used since 1988. The Vegan Society's history says it was adapted from 1944 to 1948 – meaning there was no definition of veganism for the first five years of the Society!

As veganism has become more popular, it is quite likely that the definition will continue to change and modify to reflect the many different beliefs and practices of vegans.

Debates over the definition of veganism and its core values have always been an important part of veganism. These have led to sometimes vehement disagreements over what *counts* as veganism, and the best paths forward for both understanding veganism and practising it, especially when it comes to activism. While debates over the definition of veganism, and its histories, are important to properly understanding its lineages, when thinking about what veganism is *for*, the contemporary moment is less about understanding what it is, and instead understanding what it can *do*. What veganism has always advocated is a refusal to consume animal-based products and the development of new ways of being in the world through ethics, technology and practice. While definitions of veganism are important, it is also easy to become *too* focused on them. Perhaps the reason that veganism invokes such a strength of feeling is that it is not simply about *what we eat*, but is an integral identity to many people.

Food is a cultural value and the entangling of identity with food is present in most diets (see Chapter 5). What, then, sets veganism apart is its way of engaging with and transforming the world; as cultural theorist Eva Giraud has argued, veganism is *more than a diet* and when we talk about veganism, holding onto its complexities and nuances is vitally important.

* * *

Veganism challenges what we consume, how food (and other animal commodities) are produced, and the broader relationships that humans have with the non-human world. However, vegan diets are regularly challenged as unhealthy, lacking in protein or nutrients. Similarly, questions over commodity sustainability are often raised, such as the supposedly insatiable vegan demand for avocados and quinoa (see Box 3.1). Even the benefits for animals have elicited accusations that veganism will erase cows and chickens from the Earth. In this book, I explore questions like: how strong is the evidence to back up vegan claims of improving the welfare of animals, the environment and human health? Are vegans destroying the world because they can't stop drinking almond milk? Do vegans have a protein deficiency that is making them ill? And will cows disappear from the planet as veganism grows? Spoiler: the answer to those last three is no!

In Chapters 2–4, I explore the three motivations beyond veganism – animals, environment and health – and explain why they are important to veganism, how veganism incorporates or works towards the betterment of animals, health and environment, and, crucially, explore how veganism has become established and defines itself – which isn't as simple as it might at first seem. Chapters 5–7 then look at the future trajectories of veganism through culture, technology and justice. I show how veganism is more than simply eating plants: it is deeply connected with culture and identity, it has implications for the development of technologies and it intersects with other social and political movements. Ultimately, this book looks at the visions that veganism advocates – for humans, animals and the planet – arguing that veganism offers rich and diverse pathways to a more equitable, sustainable and healthy future.

2
ANIMALS

Over the last two centuries, animal-derived products have taken up an ever-greater proportion of the human diet. But animal exploitation isn't only found in our diet. Animal products are everywhere and some of them are unavoidable: isinglass from fish is used in brewing beer; toilet paper contains gelatine from cows' bones; sugar is sometimes refined with burnt animal bones; latex condoms contain cows' milk proteins; toothpaste uses animal fats to stop it drying out; printing inks (and therefore books) contain animal fats, bones and crushed beetles; and bank notes contain tallow, an animal fat. While vegans strive to avoid animal products, boycotting ink and money and taking our own vegan toilet paper everywhere with us is, for most people, not possible or practicable. In this chapter, I explore why vegans refuse to participate in the consumption of animals, and consider how the world might transform if animals were no longer commodities, but instead co-habitants of the planet.

Farmed for food

Today, the world's biggest consumer meat market is China, where in 2021 7 per cent of the world's meat and half of all pigs were consumed.[1] Meat has become a growing staple of China's national diet, but it's worth mentioning that the per capita meat consumption is still higher in the United States at 151.4kg per person per year. Since 1982, meat consumption has increased in China by 50kg per person per year, with a further 30kg expected by 2030.[2] This increased consumption has huge implications for the planet, adding 1.8 gigatonnes of greenhouse gases to the atmosphere between 2018 and 2030,[3] contributing to increased risk of disease outbreak, a fear realized in the African swine fever outbreak in 2021, and, of concern to this chapter, for the lives and deaths of animals, most recently in the opening of high-rise multistorey pig factory farms. But the gruesome reality is that factory farming exists across the world and it is violent in all its forms (see Figure 2.1).

The conditions of animal lives and deaths in agriculture are devastating. While minimum welfare standards have been introduced in many countries, the development of new technologies for farming have led to increasingly intensive and invasive animal practices. Farming is not the only concern for vegans, and it is in fact its extension beyond farming that separates it from some other forms of animal advocacy or vegetarianism, but nonetheless veganism is overwhelmingly concerned with animal agriculture, because this is where huge numbers of animals suffer and are killed every day. There are (at least) three factors which make a focus on animals farmed for food so central to veganism:

1. the direct killing and suffering of farmed animals;
2. the knock-on consequences of animal agriculture on wild species; and
3. the (in)visibility of this suffering.

Figure 2.1: Meat production by region, 2023

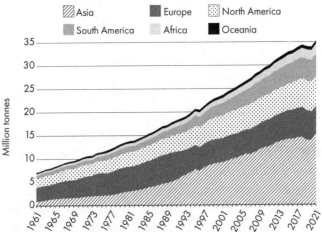

This graph, from *Our World in Data*, shows the rapid growth in meat production between 1961 and 2021. The production of meat has quadrupled over the last 60 years, and is predicted to continue rising.

The issues of intensifying animal agriculture systems have been raised for decades, with one of the earliest comprehensive accounts written by the British campaigner and author of *Animal Machines*, Ruth Harrison, in 1964. But in the intervening 60 years, the consequences of intensive farming warned about – a warming climate, more severe weather, soil degradation and loss of wild lands – have come to fruition, with disastrous environmental effects, health risks and unimaginable suffering for animals, particularly for species like cows, chickens and pigs.

Cows raised for beef have had their lives and environments transformed over the last century in the quest for efficiency and more recently to reduce their emissions. Beef has long been popular in many diets around the world, forming part of the identity of cultures such as the Latino cowboy masculinity of Texas or the barbecue culture of Australia's 'beef capital', Rockhampton, that increasingly rely on intensive farming. For example, Botswana has moved from subsistence beef production to more intensive forms. Despite the global focus on reducing emissions, ruminant consumption is expected to rise 88 per cent between 2010 and 2050. To meet this demand and appeal to sustainable goals, new technologies, particularly genomic intervention,[4] aim to produce bigger cows who live – and therefore belch methane – for less time while providing more meat. This supposed efficiency also ticks boxes for emissions reductions, playing into policy demands to act on carbon.

In his book *Porkopolis*, anthropologist Alex Blanchette recounts his experiences of working in a pig farm to understand the relationship between pigs, human workers and labour in contemporary American factory farming. What is particularly striking about his account is not just the awful conditions for pigs, but the insight into the lives of the people raising, breeding and killing pigs for meat. The arrival of the factory farm in rural areas quickly takes over the community, providing jobs in exchange for the identity of the place and the control of human workers. In practice, the workers on factory farms see their villages and towns become defined by this identity and feel their own bodies reshaped by the repetitive physical work of rearing and slaughtering animals. The factory farm is a place of exploitation for *all* animals – including humans. The work of killing animals isn't equally distributed: it's dirty, and often undertaken by low-paid, working-class and migrant labourers. As detailed in Chapter 4, it also has psychological and physical repercussions for workers.

The intensification of farming has not just harmed animals and humans by enclosure, but has also found new ways to make animal production more efficient, which has come with increased suffering of animals. The broiler chicken business, for example, has bred birds to grow bigger and faster, negatively impacting both their welfare and health. In 2018, a group of geologists led by Carys Bennett at Leicester University published a paper that argued that the broiler chicken is a new morphospecies, their biology and physiology so altered from their ancestor *G. gallus domesticus* that they can now be considered as a signal of a human-reconfigured biosphere. This transformation resulted from genomic, environmental and nutritional interventions that forced chickens to grow as big as they could, as quickly as they could, to push down the cost of meat made from their bodies. The result: broiler chickens now grow much bigger and faster than they did 50 years ago, through regimented overfeeding diets, control of movement to avoid burning calories and selective breeding. The conditions for these chickens are horrific, and they suffer from behavioural, psychological and physical ailments as well as being a hotbed for zoonotic disease and cause of ecological damage. At just eight to 12 weeks old, chickens, often bald from feather pecking, with damaged beaks, blind from the ammonia built up from their urine, are slaughtered.

The conscription of humans and animals into animal agricultural systems has links with exploitation, illness and danger at work, as well as being infused with racial, class and nationality dynamics which creates physical and psychological trauma for humans. The intensification and expansion of animal farming do not just have impacts for farmed animals, but also for wild species whose habitats are destroyed for grazing or growing crops to feed farmed animals, for workers in the sector and for the relationships between species and ecologies. For a growing number of people, not only vegans, industrial animal production is a grim reality in need of fundamental overhaul, even for those who do choose to eat animals. Industrial farming is often the target of vegan critique, due to the scale and intensity of suffering inflicted on animals, but a pivot to smaller, higher-welfare animal farming is not going to end this suffering, not least because it still ends in the slaughterhouse.

Disappearing animals

Changes in the global human diet and farming practices have led to an increase in the consumption of animal products that has coincided with many animals disappearing from human society. In the 18th century, cities such as London were filled with chickens, pigs and cows who roamed the city streets but, over the next century, cities in Europe and North America were 'sanitized' to remove animals – especially farmed animals – from them. Today, seeing many of these animals in a city in the West would be bizarre. As animals have increased in our diets, their physical presence in our lives has diminished. In fact, outside of the animals we know as pets, many people have very limited interactions with animals. One study found that while children hold non-native animals like lions and zebras in high regard, they are disconnected from and less interested in local or native species.[5]

The disappearance of animals from human spaces is closely related to the intensification of animal farming as the appetite for meat, and the ability to produce it cheaply, has grown. The enclosure of animals in industrial farms has hidden the dirty and unpleasant work of raising and killing animals. Today's intensive system is damaging not only to animals, but also to the planet through deforestation, pollution and greenhouse gas emissions, and to human health through carcinogens in meat and the risk of pandemics (see Chapters 3 and 4). However, when it comes to eating, it has been a case of 'out of sight, out of mind' for the impact of meat consumption on animals and the environment. While the provenance of food has always been important to some sections of society, questions of where it came from and how it got there have, broadly, been less important than food being available and cheap. With environmental crises leading ever more urgently to increased scrutiny of human diets, veganism has become part of a mainstream conversation on how we consume. In this mainstreaming, the centrality of animal suffering, rights and liberation have, some worry,[6] become less of a focus, in turn reinforcing veganism as purely a dietary choice, rather than a set of beliefs and practices.

Veganism as an organized ethical and political movement has always had animals at its heart, and while concern for health and the environment in vegan discourses today also have important histories, they have, in the past, been less central to veganism itself. This focus on animals has meant that veganism isn't *only* about eating, with vegan ethics extending to animals in medicine, entertainment and in the wild. However, a prioritization of the animals we eat has occurred in veganism for two reasons: first, the huge and necessary role that eating plays in human lives, and its impacts on non-humans and the environment; and, second, the food sector is the site of the most and worst animal exploitation and suffering. Animal-centred veganism is about fostering new relationships with others, both human and non-human, and challenging the most extreme spaces of violence (such as animals in food production) is foundational for widespread social and political changes.

Animal advocacy

There are many kinds of animal advocacy, some of which focus on specific species or particular campaigns (such as the ending of battery-cage chicken farming); while vegetarianism avoids any products brought about by animal slaughter, which are primarily meat products and some clothing items such as fur and leather. These kinds of advocacy for animals don't take a holistic view of animal suffering, while veganism is usually focused on all kinds of animal suffering. Veganism's relationship with and advocacy for animals therefore extends beyond farming for food and clothing, and seeks to avoid any products or spaces (such as zoos) that rely on or create animal suffering. However, there are still differences in the approaches that vegans take to animals, and their rights and welfare, that make for a diverse and lively umbrella of activism.

Veganism is often talked about in the same breath as animal rights, but they aren't synonymous. Rather, they are overlapping movements with different aims and visions at their heart. Action on behalf of animals has existed throughout human history, with particularly long roots in Eastern religions, such as in the idea of *ahimsa* (nonviolence) in Buddhism, a concept taken up by the American Vegan Society's Indian-Persian American founder Jay Dinshah to attempt to bring a different ethical shape to veganism from 1960 to his death in 2000. Although Eastern religious influences have not become dominant in Western veganism, inflections of spirituality are commonplace in animal advocacy, attesting to these roots (see Chapter 5 on cultural identities). The organized animal rights movement, like veganism, has also been described as originating in Britain, with some tracing the idea that animals are beings, not resources, back to the founding of the Royal Society for the Prevention of Cruelty to Animals (RSPCA) in London in 1824, although many would rightly see the RSPCA today as a welfare organization, primarily focused on 'pet' animals, and not interested in rights. Nonetheless, the RSPCA was the first known animal advocacy organization, and its early emphasis was improving the lives of working animals, especially horses, in London.

Animal rights are rooted in legal and moral frameworks of justice that aim to understand, interpret and end suffering and pain for animals. In many accounts of veganism and animal rights, there are two figures seen as the 'godfathers' of the modern animal movement, inspiring both theoretical work and activism: philosophers Peter Singer and Tom Regan. Singer, in particular, is widely credited as the founder of the contemporary animal activist movement, after the publication of *Animal Liberation* in 1975, and is consistently cited by academics and animal advocates as initiating both the academic and activist political wings of animal activism.[7] Regan is often regarded as Singer's counterpart, with his deontological ethics (or moral obligations) set out in *The Case for Animal Rights* (1983), which offers an abolitionist perspective on eradicating animal use through the idea of all animals being deserving of respect, care and freedom from suffering. Contrary to popular belief, it was only Regan who believed in moral rights for animals as inherent, not something given, while Singer is more concerned with minimizing suffering as part of his utilitarian commitments and avoids rights-based approaches. Approaches to animal rights were neither begun nor finished with these so-called fathers of theory. Indeed, Singer himself has credited his inspiration for thinking about animals to Roslind Godlovitch, whose philosophical arguments convinced him that 'the logic of the vegetarian position was irrefutable'.[8] The intellectual tradition of animal rights has been widely transposed onto veganism and into vegan ethics, despite veganism's intellectual and ethical foundations not always aligning with the ideas of animal rights. An animal having rights does not necessarily mean that humans cannot use, or eat, them, subject to particular conditions. For vegans, this is never acceptable.

Animal ethics have been rooted in multiple heritages, some in animal rights, others in religious and cultural histories, but also in traditions that bring together animals and humans. One of the most developed and compelling of these is ecofeminism, which offers both intellectual and practical arguments for veganism. In 1990, Carol J. Adams published *The Sexual Politics of Meat*, a book that has gone on to reshape and transform the conversation around meat-eating, gender and politics. Her work was part of an ecology of ecofeminist work which addresses how 'the various ways that sexism, hetero-normativity, racism, colonialism, and ableism are informed by and support speciesism and [analyses] how the ways these forces intersect can produce less violent, more just practices'.[9] While Singer and Regan are often, in my opinion incorrectly and ahistorically, credited as the fathers of the animal movement, there were countless animal movements – usually led by women – operating throughout the 19th and 20th centuries. While those women were often dismissed as 'sentimental', their focus was often on connecting the fight for animals with human justice, arguing and acting for the causes together. Often, these links between human and animal movements prioritized emotional politics and practices, not just rational arguments. Today, these fights continue with a focus in overlapping feminist and vegan activism on farmed animals and the control of reproduction for both humans and non-humans (as I discuss further in Chapter 7).

Animal byproducts

Veganism's ethical and political case is strongest when it comes to animals. Vegans' motivations to end animal exploitation aren't individual reactions, but an engaged and politically motivated movement to change societal structures. However, the growing popularity and visibility of veganism have led to it being cast as a lifestyle or diet that is malleable or extreme. These critiques of veganism as an individual dietary or consumption choice ignore the collectivist and visionary organization of veganism. In contrast, meat reduction or flexitarianism are increasingly being touted as the way to save the world, but as they don't represent any significant change in our relationship with animals, they don't offer a cohesive plan for change. A focus on reducing the amount of intensively farmed meat consumed, for example, can obscure the violence of animal 'byproducts' (such as eggs, dairy and honey), which, despite the name, are anything but byproducts.

Eggs, dairy and honey have huge industries of their own, producing unique kinds of animal suffering, that have little or nothing to do with the animal meat sector. Honey, for example, is often seen as a relatively benign foodstuff, that does not harm the bees who produce it; while the collection of honey may not *hurt* bees, it does exploit them and worsen their health, as the honey they would usually consume is replaced with sugar substitutes which don't provide the nutrients they need. Contemporary beekeeping seeks to increase the efficiency and production of hives through selective breeding, increasing disease in populations, and bees are often abandoned or killed once they are no longer productive enough. Despite beekeeping being linked with sustainable ways of life, the increased numbers of farmed bees are negatively impacting wild bee populations who are so crucial to pollination.

The biggest industries for animal byproducts are the dairy and egg industries (see Box 2.1), demand for which has shifted in recent decades as both chicken and cow meat has increased. Far from being byproducts of meat production, these are their own production ecosystems. India is the largest milk producer in the world, consuming almost all of the 192 megatons[10] of dairy it produces annually domestically, but new markets are emerging for consumption and trade, notably in China. Chicken egg consumption dropped in the wake of salmonella outbreaks in the 1990s, with the market shrinking in North America, despite the rapid increase in consumption of chicken meat. In 2022, the price of eggs increased drastically as a particularly bad outbreak of avian influenza led to millions of chickens worldwide being culled, on top of soaring feed prices due to supply chain shocks. Yet, it is only vegans that act on the ethical problems with consuming these food products under any circumstances.

Box 2.1: Egg and dairy production

Both the egg and dairy industries are functionally separate from chicken meat and beef production, with the breeds of animals, their conditions and their treatment all being specifically tailored to producing high quantities of milk and eggs. Laying chickens are generally much smaller than meat birds and produce twice as many eggs due to selective breeding over the last century. They have also been bred to produce larger eggs to appeal to consumers, which means laying hens suffer from egg-related diseases and health issues, such as egg binding (when an egg is too large to be laid) and egg yolk peritonitis (the formation of yolk in the abdomen which can build up and lead to bacterial infections, which is the leading cause of death in commercial flocks).

Similar breeding processes and genetic selection have been put to work in the dairy sector: Holstein-Friesian cows (large white and black cows) have become the most common breeds in the United Kingdom, Europe and the United States due to their high milk yields, increased over four centuries of selective breeding, while the Gir cow (large, indigenous brown cows) is the most popular dairy cow in India due to their long life-span and ability to have 10–12 calves, thus extending their productive lives. These breeding processes have led to higher milk yields, but also increased prevalence of lameness, mastitis and metabolic disease.

In addition to suffering through increased illness, the conditions of both egg and dairy animals are no better than those of meat animals and have their own specific violences – such as the slaughter of male animals considered 'surplus' to the industry. While in the dairy industry, the immediate shooting of male calves is becoming less common (with them instead being raised for beef or veal), in the egg industry, the killing of male chicks is usually done through extremely painful processes of gassing, suffocation, electrocution or maceration (high-speed grinding, which chicks go into live).

Eggs and dairy have been described as forms of feminized protein, 'produced through the abuse of the reproductive cycle of female animals' manipulated for human needs.[11] In my conversations with vegans over the last ten years, people have frequently pointed to the ongoing, intimate violences of dairy production as the most objectionable of all kinds of food production – even more so than meat production – due to its repetitive and extended physical and psychological cruelty, that ultimately also ends in the slaughterhouse. To produce dairy, cows are artificially inseminated by what is colloquially referred to by farmers as 'the rape rack', carry their calves for nine months, and soon after birthing, are separated from their calves to ensure that the mothers can be milked for humans. Calves often go on to live in isolated crates where they are fattened for a kind of meat known as veal. Dairy cows are then milked daily on huge pieces of machinery akin to conveyor belts. This process was heartbreakingly represented in Andrea Arnold's 2022 documentary, *Cow*, which followed the life of a dairy cow, Luma, on a British farm after the birth of her sixth calf, and dispelled any lingering myths of happy cows in rolling fields of green.

Human–animal connections

While there are various reasons a person may become vegetarian rather than vegan (such as taste or preference), the shared commitment to avoiding animal meat often leads to the conflation of the two as similar ethical stances. However, vegetarianism and veganism differ greatly in the way they conceptualize animal care and cruelty, with veganism offering a more expansive idea of what 'counts' as suffering for animals, in turn challenging the idea that animals should be used by humans at all. Veganism isn't, therefore, just about *not* eating animals: it is about rethinking the human relationship with the non-human world through the lens of our consumption, but also through our treatment of other species, environments and humans. As such, it expands across the social, political, economic and ethical spheres, to refuse violence and promote compassion.

While there is a focus on eating in contemporary discourses around veganism, the exploitation of animals is insidious and extends far beyond the food system, examples of which I detailed at the opening of this chapter. There's little doubt that in recent years, veganism has grown in prominence in popular culture, contributing to changing public discourses around how we treat other animals. While this has been linked to an increased interest in sustainable and healthy eating, there has also been mounting evidence of societal shifts in how humans think about our relationships with animals and the natural world. During the COVID-19 pandemic, when many people were spending a lot more time at home than usual and access to the outdoors was limited, the lessened movements and presence of humans allowed animals to roam and recolonize some urban areas, and it also enabled people to find more value in watching or spending time with other species.

The plight and enjoyment of these wild animals is also part of a vegan ethics.

Scientists have argued that we are in the midst of the Sixth Mass Extinction of species and that this is the direct result of human activities. The United Nations and other organizations have attributed a great deal of habitat loss to the food system, notably the destruction of the Amazon rainforest for soy to feed cattle, and directly for cattle ranching. Overfishing has led to the declining health of oceans and seas, which, combined with a plastic crisis, has polluted and acidified waters leading to fish, mammal and coral loss. Meanwhile, the increased use of pesticides has had disastrous consequences for insects, birds, plants and humans. Yet, the direct connection between animal agriculture and wild species loss is often difficult to visualize.

Photojournalist and animal activist Jo-Anne McArthur has dedicated her career to documenting the lives and deaths of exploited animals and advocating for their rights. Her work has exposed animal suffering in factory farms, circuses and protest, but also showcased sanctuaries and rescues across the world. One of her most famous and moving pictures, 'Hope in a burned plantation', was taken in Victoria during the 2020 Australian wildfires (see Figure 2.2). In the centre of the frame is a kangaroo with a joey in her pouch. The kangaroo stares directly into the lens, alert, while her joey looks out. Behind them, the forest glows orange, and beneath their feet is scorched black earth.[12] The raging of wildfires is directly connected to the impact of industrial agriculture on global warming,

Figure 2.2: 'Hope in a burned plantation', 2020

This photograph of a kangaroo and her joey was taken by Jo-Anne McArthur during the Australian wildfires in 2020. McArthur documented the effects of the fires on both wild and domestic animals, as approximately three billion animals were killed or displaced. In 2021, the photograph was Highly Commended in the Wildlife Photographer of the Year's People's Choice award, voted for by the public.

and this image represents how this kangaroo is no less part of a vegan ethics than a cow or sheep, even if they are less common in mainstream conversations around veganism.

Looking closer to home, vegans have also critiqued the relationships between humans and the animals that are kept as 'pets'. Dogs, cats and other creatures who humans commonly live with in their homes occupy a special space in society and are valued more highly than other animals. However, these relationships are deeply unequal, with domination and power being exerted over animals to control their territories, behaviours and freedoms. Pedigree dogs are notorious for suffering from ill health due to their breeding, while there have been increasing concerns raised over the impact of cats on local biodiversity and nature. But the right to keep animals as pets is one that is fiercely defended, including by many vegans, although a more radical thread of veganism argues that pet ownership is an unequal relationship and one that we should abolish. Concerns over pets aren't restricted to vegans: the overflowing shelters for pet animals are often in the news, and there are resurging debates over what kinds of animals it is acceptable to keep in domestic settings. But the idea that we shouldn't be keeping pets *at all* is a radical one that challenges how people think of themselves. My own opinion on living with animals in our homes, based on my experiences of rescued guinea pigs and chickens becoming part of my family, is that we should look to home animals as equals, and that people who live with animals must be committed to education about animals' physical and emotional needs, adopting and being lifelong providers for the animals they live with. While animals in the home don't face the same violence as farmed animals, neglect and a lack of understanding about their needs means many 'pet' animals do not have good lives, even if their basic needs are met. However, there is heated debate within and outside of the vegan community on what a future with animals looks like.

Concerns over wild and companion animals are not limited to vegans; charities the world over are dedicated to raising awareness of and funding for species like dogs, cats, horses, orangutans, pandas, polar bears and so on. However, there is a hierarchy of species in society, and advocacy: companion animals are beloved and hurting them evokes anger, while the suffering of farmed animals has also led to welfare reforms. Rare and endangered charismatic species, like whales, elephants or lions (at least theoretically) have legal protections. But most animals on the planet don't fit into any of these categories: insects, fish and small mammals have their advocates, but they are much further down political agendas of rescue and protection. Many of them aren't even legally recognized as sentient, meaning there is little to no oversight of their treatment in farming or pharmaceutical industries. While it is not only vegans, then, who care about or advocate for animals, veganism challenges the hierarchy of species, seeking for concern to be extended to *all* animals. Ultimately, veganism as an animal-focused ethics and practice aims to change our relationships not just with the animals we eat, but with those we live with, love, hate and even ignore.

* * *

Veganism 'for the animals' aims to find new ways of relating to animals and, therefore, also to society, culture, the environment and other human and non-human beings. A vision for a future that centres animals as equals is one that is radical in most societies today, and it is therefore met with resistance, mockery and dismissal, at both policy and interpersonal levels. With the future of the planet on the line, veganism as a dietary intervention has started to be taken a bit more seriously, with plant-based products becoming much more common but, in the process, animals themselves have faded into the background as veganism and plant-based eating have increasingly become conflated. Nonetheless, animals have been and will continue to be central to veganism, despite an increased focus on the environmental and health-based cases for veganism.

Before moving on to health and the environment, though, I want to reflect on a question often posed to vegans: *what would happen to all the farmed animals if we all went vegan?* This question can be frustrating for vegans. For one thing, it is incredibly unlikely to happen. It is also often posed by devil's advocates wanting to undermine or dismiss veganism. But I want to take it seriously.

If the whole world went vegan, it wouldn't happen overnight. It probably wouldn't come from individual change either, but a combination of changed ethics and values at social and political levels, in combination with policy intervention and structural transformation to support dietary transition. Over this long process, farmers could be supported to transition, education on nutrition could be circulated and animals themselves could stop being forcibly bred – because farmed animals aren't reproducing without human intervention. Through the years – perhaps even decades – of this transition, the number of farmed animals would naturally drop; the survivors of the animal agriculture industry could be transitioned into sanctuaries to live out their retirements. This would be supported because if the whole world transitioned to veganism, it couldn't be by force; it would require a transformed relationship with animals where people would *want* to protect, love and dedicate space to them. This multispecies utopia would be the result of a world that went vegan, so cows, dogs, chickens, insects and fish wouldn't be abandoned but would become full and equal members of social and political life.

Of course, this speculative future almost certainly won't come to pass, even as veganism grows, although there are spaces and sanctuaries where something like this already exists. But I believe it is worth emphasizing that veganism through an animal lens isn't just a politics of refusal, it is a practice and dedication to hope for the future. Veganism, in this form, is a way of life that encompasses empathy, care and justice, and thus imagines and practices new kinds of relationships with the human and non-human world. A veganism for the animals, then, is just that: *for* the animals, but for us humans as well.

3
ENVIRONMENT

While vegan practices have long had animals at their heart, veganism's recent growth can be attributed, at least in part, to increased engagement with environmental causes. It is no secret that food systems are broken, for people and for the planet: during COVID-19, the fragility of supply chains due to bottlenecks in labour, processing and transport was visible on empty supermarket shelves. The Russian invasion of Ukraine in 2022 created a drastic drop in exports of wheat and corn that previously produced 10 per cent of the global market. These shocks to food systems come on the heels of ongoing crises related to environmental degradation of land and a warming climate.

The global food system is based on a model of extraction and colonial expansion that is unsustainable. The effects of the food system on the planet are well-documented, as both contributors to climate change and vulnerable to its most severe consequences (see

Figure 3.1). Over the last millennium, three-quarters of the Earth's land surface has been changed by human use, with the period since 1960 seeing a net loss of almost a million km^2 of forest to agriculture, concentrated in the Brazilian Amazon and rural China.[1] Within this context of a warming, more volatile and degraded planet, the rise of veganism as an environmental intervention raises the question: how can veganism improve or lessen the impact of humans on the environment, and what are its commitments to doing so?

Red warning for humanity

In 2006, the United Nation's climate change body reported that 'livestock [is] one of the top two or three most significant contributors to environmental problems, at every scale from local to global'.[2] Over the next 15 years, there was little to no action to reduce livestock production globally and, on average, people are eating more meat today than ever before. In 2021, UN Secretary-General António Guterres issued a code red, saying 'the alarm bells are deafening, and the evidence is irrefutable: greenhouse gas emissions from fossil fuel burning and deforestation are choking our planet and putting billions of people at immediate risk'.[3] Notably, there is no mention of the role industrial animal agriculture has played in this, which is perhaps unsurprising given that, in debates around climate change mitigation and adaptation, the food system is an issue that raises intense disagreement. This

Figure 3.1: Greenhouse gas emissions per kilogram of food product, 2023

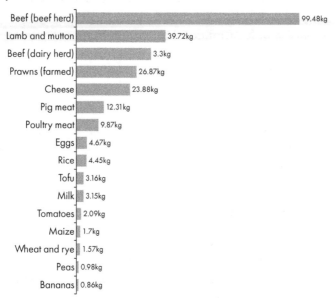

Beef (beef herd)	99.48kg
Lamb and mutton	39.72kg
Beef (dairy herd)	3.3kg
Prawns (farmed)	26.87kg
Cheese	23.88kg
Pig meat	12.31kg
Poultry meat	9.87kg
Eggs	4.67kg
Rice	4.45kg
Tofu	3.16kg
Milk	3.15kg
Tomatoes	2.09kg
Maize	1.7kg
Wheat and rye	1.57kg
Peas	0.98kg
Bananas	0.86kg

This graph from *Our World in Data* is based on Joseph Poore and Thomas Nemecek's 2018 research on greenhouse gas emissions per kilogram of food product. Beef eclipses all other food types, and there is a clear division in carbon emissions from animal and plant products.

has led to agriculture being off the agenda altogether; at the global climate conference COP26 in Glasgow, for example, there was little talk about emissions from farming, and the menu was full of beef and dairy, which led to protests by climate activists.

Nonetheless, people concerned with the environment know the impacts of their dietary choices, and the rise of environmentally conscious eating seeks to address the central problem of global, national and local food systems. For example, locavorism and seasonal eating appeal to those seeking to reduce the air miles that their food travels, while recently popularized regenerative farming claims to merge agriculture with conservation, although the environmental benefits of these methods are questionable.[4] Within different practices of environmentally conscious eating, veganism is often painted as an outlier – dismissed as an unserious and impossible option, and one that primarily works as an individual consumption choice, despite growing evidence that vegan environmental ethics are practicable and could be an effective environmental strategy long-term.

Environmental crisis

Animal agriculture is having a devastating environmental effect. Soil degradation is highest in animal agricultural land and, with 26 per cent of the Earth's ice-free land being used for livestock grazing, this is no small issue. In the United States alone, animal agriculture is responsible for 85 per cent of all soil erosion.[5] And it isn't just impacting land: pig, poultry and cow farming produces massive amounts of waste that end up in oceans and factory farming is the leading cause of ocean dead zones. Waste from intensive fish farms pollutes the water, impacting surrounding marine life, and salmon farming caused about $50 billion of environmental damage between 2013 and 2019 through pollution and parasites.[6]

The air and atmosphere are also not safe from agricultural methane emissions. There are also lagoons of animal waste that emit ammonia – a pungent gas that affects soil, but also human respiratory systems and eyes. In the United Kingdom, ammonia is damaging 60 per cent of land, with knock-on effects for humans, animals, plants and entire ecosystems.[7]

Attempts to precisely rank the causes of environmental and climate crises are futile. Totalling up emissions, financial or natural losses, or weighing one environmental ill against another, is almost impossible, especially when accounting for complex social and economic systems. Almost everything humans do across the world today damages the environment, whether that is industrial production, air travel, heating and electricity, agriculture, transportation or work. To understand how veganism has become a part of these debates, it's important to be aware of the state of the planet, and how the dire situation we are in has been shaped by extractive relationships that deplete animals, land and plants by treating them merely as resources, without any interest in renewal or sustainability.

In 2009, the Stockholm Resilience Centre defined nine planetary boundaries within which humanity can develop and thrive, but which, if breached, will make large-scale, abrupt and irreversible changes to the Earth's environment more likely. By 2015, researchers had evidence that several of these planetary boundaries (climate change, biodiversity loss, shifts in nutrient cycles and land use) had been crossed as a result of human activities, putting the Earth system in unprecedented territory. In 2022, the same scientists announced that the environmental pollutants boundary had also been crossed, which means that the use of pesticides, antibiotics, plastics and industrial chemicals now exceed levels safe for human and non-human habitation, with particular impacts on the atmosphere and soil. The breaching of these boundaries is due to the entirety of human activity, but animal-based food systems are part of this catastrophe, something the Stockholm Resilience Centre has strongly emphasized. Land use, biodiversity loss, nutrient cycles and pesticides and antibiotic use are directly connected to intensive animal agriculture. The intensification of animal agriculture is part of a broader system of human domination over nature.

Our contemporary relationship with nature sees the world, including animals, as a vast resource for profit, and has long ignored the fact that nature can only be exploited so far without doing irreparable damage – regardless of the ethics of this relationship. Over the past 30 years, global chicken farming has seen exponential growth, but despite claims of chicken consumption being 'greener' than beef, poultry farms are responsible for significant environmental degradation. In Britain, the River Wye has been flooded with phosphorus from intensive chicken farming, leading to a dying and polluted river with dwindling biodiversity. In the United States, where chicken is taking up an ever-increasing proportion of the national diet, carbon emissions from chicken farming are estimated to equal to that of running

12.37 million cars.[8] Supposed 'circular' economies of agriculture, such as the use of chicken waste as fertilizer, are also environmentally dubious. Chicken waste fertilizer impacts the phosphorus and nitrogen cycles of the soil, leading to the creation of dead zones in rivers. While chickens might have lower direct emissions from their agriculture than beef production, this is a misrepresentation of a complex system, where chicken farming has a significant environmental impact, chemical and microbiological risks and health consequences for workers and communities around intensive farms (see Chapter 4). Comparisons of carbon emissions are only part of the story.

Problems of consumption

The consumption of animals is expected to continue rising, notably in Africa and Asia, where the spread and further intensification of animal agriculture are accompanied by looser regulations, increasing environmental and public health risks. High-rise pig farms in China, intensive cattle feedlots in the United States and chicken megafarms in the United Kingdom are being opened on the grounds of efficiency, but also for allegedly environmental reasons, as intensive systems reduce the need for deforestation. Any system of food production has to weigh up its environmental costs and welfare commitments to feed a growing population. Demand for different kinds of food and production systems vary across the world, although over the past century, food systems have in general unified to aim for greater efficiency and increased uniformity.

Global food systems today are more connected and larger than at any point in history, with longer supply chains, due to the increased accessibility of transportation and refrigeration. With this interconnectedness, food has become cheaper, and availability is no longer seasonal or even regional, which has led to new patterns of consumption. The demand for more foods to consume has led to environmental degradation and deforestation, as well as the exploitation of human labour; both animal and crop agriculture today are perfect storms for modern-day slavery, with an estimated 10,000 to 13,000 people being exploited in the United Kingdom alone.[9] Contemporary global food systems have severe consequences for humans, animals and the environment across the world, and these are not disconnected from one another. These problems are related to the ways we, as humans, consume.

As noted earlier in this chapter, in China, meat consumption is on the rise, but China is also the heart of new food product innovation, as it plans to grow an export market for alternative proteins as well as feeding a large post-industrializing population with increasing wealth. As population grows, food consumption grows, obviously, but not only does the total amount of meat consumed across a population grow as there are more people eating, but the amount of meat consumed by individuals also increases.[10] With social, economic and political change, the *kinds* of food that people eat also change. In many countries, eating meat is aspirational and a sign of affluence, much like it was in North America and Europe in the 19th and 20th centuries. However, this has never been true everywhere; for example, in India, vegetarianism is an upper-caste signifier related to Hinduism, and meat-eating is socially and politically stigmatized and bound up with the oppression of Muslims, for whom beef-eating has become a political act of resistance. Increased meat consumption, seen most steeply in China, Asia, Central and South America and Africa as their meat consumption catches up with the West, has significant environmental impacts. Rising populations are emphatically *not* the problem, but associated consumption changes do have environmental impacts.

Environmental problems aren't confined to one country, or even just to food. The consumption of almost every nation on Earth is unsustainable: if the whole world lived like the United States, we would need the resources of five planets; for China or the United Kingdom, we would need four. While veganism offers one solution to some of these environmental problems, it is not currently being pursued via policy at national or international level anywhere. Instead, a focus on growth, progress and animal-based food production continues to dominate (and, as discussed in Chapter 6, this extends to new food technologies). This ignores the specific damage that animal consumption has had on the planet, and the potential that exists for different modes of consumption – and production – that are less environmentally devastating.

With growing concerns over carbon emissions and the environmental impacts of animal-based food systems, the time is ripe for plant-based food systems to offer an alternative.

Vegan environmentalism

In the last half-century, environmentalism has been deeply entwined with veganism. In recent years, the urgency of mitigating climate change has catalysed veganism's growth. However, within environmentally progressive politics, veganism is still not seen as a viable large-scale mitigation strategy, despite the evidence that plant-based foods are much more environmentally friendly than even the lowest-impact animal foods.[11] Veganism is gaining some acceptance within the wider green movement but is still not fully integrated. For example, the creation of Animal Rebellion as an offshoot of Extinction Rebellion has played out some of these problematic relationships, with Animal Rebellion (now Animal Rising) announcing in 2023 that it would part ways with the environmental group over concerns that the wider environmental movement continued to ignore the importance of animal liberation and its role in reconnecting with the natural world.

While the vegan diet is often condemned for stereotypically relying on famously climate-unfriendly avocados and almonds, veganism in practice often goes together with organic, local and seasonal food choices. For example, veganic (vegan and organic) agriculture which is 'free of synthetic and animal-based inputs' offers promise in 'diverse realms such as food safety, environmental sustainability and animal liberation'.[12] The vegan and organic movements share a commitment to agroecological alternatives, and the veganic agriculture movement merges the two to ensure food is both synthetics- and animal-free and has had a set of standards since 2007.[13] Similarly, veganism's political ethics overlaps with other food justice movements, such as freeganism, which is an anti-consumerist form of dumpster diving for food.

Despite the good intentions and political urgency of organic, grow-your-own and DIY food production for health and wellbeing, future food systems cannot be sustained by small-scale projects alone. Anyone who has grown, or attempted to grow, their own vegetables and herbs will be aware of the hard work and care required to produce a handful of tomatoes or a bowl of potatoes that would last a few meals, and be able to imagine the impossibility of scaling this up to produce a year's worth of nutritional calories, even if they quit their jobs to focus on it full time. Growing food takes an enormous amount of time, space and energy.

The majority of contemporary food production systems operate by extracting value from animals, humans and plants without compensating for it ecologically (or financially), meaning that almost any kind of food production involves the pain or exploitation of someone or something. There is currently almost no way to produce food at a global scale without some degree of harm, which has led to countless philosophical reflections on how we

Box 3.1: Who ate all the avocados?

There isn't, of course, a simple answer to complex environmental problems. Veganism is, undoubtedly, better for the planet than eating animals, but that doesn't make it *de facto* environmentally friendly or environmentally neutral. Plant and crop farming also has environmental impacts, particularly monocrops which destroy soil and land.

Critiques of veganism throw allegations that monocrops of almonds, avocados, soy and coconuts are destroying the environment, and this is the fault of veganism. If these are indeed foods driven by vegans, then the approximately 88 million vegans across the world would be eating approximately nine million tons of avocados, 13.9 million tons of almonds, 63 million tons of coconuts and 349 million tons of soy a year.

Per person, this equates to just over 100 kilos of avocados (about 400), almost 160 kilos of almonds (about 100,000), 715 kilos of coconuts (about 1,050), and almost 4,000 kilos of soy (about 47,000 burgers with 85g of soy) *every year!* If this is true, I've not been eating my share.

Clearly, the production of these foods is not driven by the small number of vegans alone, and while shifting blame isn't going to improve the environment, it's absolutely important to point out where claims about veganism are made in bad faith.

should eat and act to reduce that harm. Technological advancement may offer new ways of ensuring harm does not come to, for example, insects in the production of crops. In the Netherlands, vertical farms (indoor, multi-level and highly controlled growing environments) are already producing entirely insect-free herb crops, which are being sold to kosher stores to provide approved 'no insect' vegetables for Jewish communities whose religious beliefs mean eating insects, even accidentally, is strictly forbidden.[14] These buildings take up less land and have a lower environmental impact, especially when combined with other emerging technologies such as hydroponics systems that grow vegetables in nutrient-enriched water instead of soil.

Despite promising technologies (more of which are discussed in Chapter 6) that might enable environmentally fairer food production, a global transition to a plant-based diet is still seen as unfeasible. But building effective food production centres is not impossible; it is already happening.

Vegan food production: a case study

The Isle of Bute, a small island off the west coast of southern Scotland, is home to a vegan food company called Bute Island Foods. Launched in 1988, Bute Island's vegan cheese substitute, *sheese*, is made from soya and, following early success, in 1994 the company renovated a building on Bute to scale up its operations and introduce seven new flavours of sheese. Demand continued to grow and the company moved into a larger premises on the island in 2013, after signing a deal with UK supermarket Tesco to sell in its stores. In the decade since, sheese has continued to increase in popularity, and the company produce products under the own-brand labels of several supermarkets and provide sheese for pizza chain Papa Johns.

Sheese is a processed cheese substitute and the production processes aren't DIY or grassroots-led or even operating in an alternative food network: the company is a business first and foremost, trying to make it within the current food system, but one with a particular kind of vegan ethics (reducing animal suffering) at its heart, rather than a radical vision of anti-capitalist production. During the company's success, the Isle of Bute has suffered from depopulation, with one in five people leaving the island between 2000 and 2021. In 2023, the vegan cheese makers are the island's largest employer and a key part of a tourism campaign to rebrand Bute as a food and drink destination. Over the last 35 years, Bute Island Foods has gone from a small food producer to a powerful economic force, shifting both the vegan food market and a local economy.

While no doubt a success story and an example for how vegan production can grow and support local economies, its more recent history raises another pressing concern and critique. In 2021, Bute Island Foods was bought by Canadian dairy giant Saputo in a multi-million-pound deal, which included a donation of almost £3 million to local Bute community initiatives.[15] But even a charitable donation can't mask the stink of a deal which means that buying a beloved vegan product is lining the pockets of a dairy production giant. This corporate capture of vegan food production has been consistently critiqued by vegans, who don't want to support animal agriculture, and non-vegans, who use the potential for corporate capture as a way to dismiss veganism as a viable ethical and political commitment. Despite the disappointment of corporate capture, and this being a growing risk as vegan foods gain popularity and make more money, this does not mean that imagining and striving towards vegan production is futile, especially at a moment when the environmentally positive potential of veganism is increasingly well-recognized.

A vegan global food system

The first Earth Day, in April 1970, captured the early days of environmental activism in the United States, according to US journalist Denise Chow, and 'sparked huge protests, teach-ins ... and rais[ed] public awareness about threats to the environment' across the political spectrum.[16] Since the 1970s, environmental awareness has steadily grown, especially over pollution, environmental degradation and the depleted ozone layer. By the late 1980s, the warming of the climate was becoming an increasing concern, with the United Nations Environment Programme's Intergovernmental Panel on Climate Change being established in 1988, and the securitization and protection of food production have been central to its work throughout its existence. Through the 1990s, projections of world food demand predicted this would begin to exceed supplies into the 21st century without biotechnological innovations, and this was without considering the impacts of climate change. As the planet has experienced the climate changing, food production and consumption are becoming an increasingly important element of planning for a warming future.

Veganism doesn't exist in a vacuum but operates in collaboration and competition with other visions of what a sustainable future might look like. While veganism has long been presented as an antagonist to the agricultural sector, its continual growth and development is envisioning new ways of producing food. Veganism is most commonly discussed by its critics as an individual consumption practice, and by vegans themselves as an ethics and politics, but if it is going to become a global force, it also needs to provide a clear vision for what vegan food production would look like. This shift from thinking about consumption to thinking about production is a crucial one in realizing the feasibility and impact of mass dietary transition.

One of the most well-cited reports on the potentials of a vegan agricultural system is Joseph Poore and Thomas Nemecek's meta-analysis of the environmental impacts of food systems.[17] They concluded that a vegan agricultural system would need just a quarter of the land that is used today. Today, less than half of the world's cereal production is fed directly to humans, with 41 per cent being used for animal feed and 11 per cent for biofuel. In Europe, less than a third of cereal is consumed by humans and in the United States, it is less than 10 per cent.[18] The production of animals as food also wastes energy and protein and is an inefficient way of producing calorific and nutritious food. In an experiment that modelled ecological outcomes from different potential future food systems, most vegan systems showed better environmental performance than our contemporary ones, especially when vegan systems were combined with organic approaches to farming.[19] The study concluded that a transition to a vegan world food system would help to protect planetary boundaries and provide sufficient nutrients.

Vegan Australia has been exploring what this vegan transition would mean for land use, environment and sustainability, economy, and health and society.[20] In Australia, animal agriculture currently makes up 1.2 per cent of the economy, provides 1.2 per cent of jobs and uses 54 per cent of Australian land, while emitting around half of Australia's greenhouse gases. Any proposed vegan agricultural system would need to maintain levels of food provision, maintain living standards, protect the environment and minimize economic and employment costs. Vegan Australia therefore quantified how much more plant-based food Australia would need to replace meat in the population's diet: 70g per person per day of animal protein, 640,000 tonnes per year, would equate to about three million tonnes of beans, chickpeas and lentils. Growing this extra food would require 0.3 per cent of Australia's landmass, plus an extra 0.2 per cent for exports. Obviously, this just addresses protein in the diet, but would reduce the land needed to produce protein from 54 per cent to just 4.5 per cent.

The land that's been freed up could then be used for rewilding, carbon sequestration, forestry, as well as additional crops for humans (see Figure 3.2). The vegan transition would also require retraining agricultural workers for the crop sector, but shouldn't lose any of the workforce, which is anyway one of the smallest contributors to the economy. This proposal would have positive environmental benefits by reducing carbon emissions and would also help land to recover. Vegan Australia provides a case study of Wooleen in Western Australia where, after one hundred years of grazing, the land was eroded and degraded. The leaseholders decided to destock all their land of animals and, after four years, were impressed that the re-establishment of vegetation had progressed better than expected, despite a drought, with threatened plant and animal species returning. This is promising evidence of the restorative ecological, as well as climatic, benefits of ending animal agriculture.

Returning to Poore and Nemecek's analysis, transition to a plant-based food system would mean that 75 per cent of the land currently used for food production could be used for other ecologically beneficial purposes. In the United Kingdom, carbon sequestration (the removal of carbon dioxide from the atmosphere) has been advocated by the Committee on Climate Change, which states the need to reallocate 22 per cent of current agricultural land to capture and store

Figure 3.2: Food consumption per capita today (top) and a proposed vegan food system (bottom)

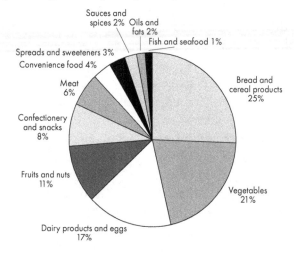

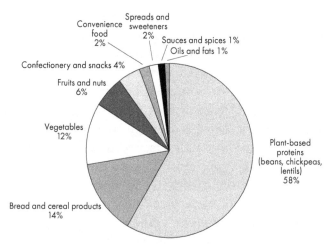

These two pie charts compare a rudimentary breakdown of food consumption per capita of key food groups today and food breakdown in a proposed vegan food system based on the need for five times more calories coming from vegetables such as beans, chickpeas and lentils.

carbon by restoring natural habitats and improving soil health.[21] In the tropics, deforestation for agriculture is devastating over half of the world's forests, and pastureland for animal agriculture is the main driver of this. Reforestation for carbon sequestration here could operate as a particularly effective carbon offset in conjunction with the transition to a vegan food system. Reallocating global agricultural land, in conjunction with remaining agricultural lands focusing on soil health, and perhaps the maintenance of animals in sanctuary spaces as grazers, would have huge planetary benefits, and there still would be a huge amount of agricultural land freed up from the transition.

About 38 per cent of global land surface is currently agricultural, and three-quarters of that is either for animals or for animal food. If 22 per cent of that animal agricultural land was reallocated to carbon sequestration, there would still be about 23 per cent of land freed up from animal agriculture which could be put to work for another environmental campaign. The 30x30 agenda was agreed as part of the UN Convention on Biological Diversity in 2022. It is one of 23 targets attempting to reverse habitat and species loss. 30x30 calls for 30 per cent of terrestrial and marine areas to be in effective protection and management by 2030. Currently, about 17 per cent of inland areas and just 8 per cent of marine areas are protected. Rewilding 13 per cent of agricultural land could easily enable the global community to achieve this target and more. Turning briefly to the sea, the total elimination of aquaculture and fishing (which currently accounts for about 17 per cent of global meat) would enable much smoother paths to legal protection for the oceans. However, about one in ten people across the world are supported economically by fishing and associated industries (such as packaging and transport). While transport and packaging could transfer their skills easily to the plant-based economy, fishers' intimate knowledge of the sea could be essential to informing where to protect and how.

Reallocating agricultural lands could enhance rather than destroy the environment, and there would still remain land freed up that could enable sustainable urban growth as the global population continues to grow. This could enable the planet to flourish, especially in collaboration with technological innovation (see Chapter 6). However, climate change is already ongoing and impacting what can be grown and where, which would affect any future food system. Rising sea levels in coastal areas may lead to a total loss of agricultural land, warmer climates mean more pests and disease and potential sites of effective agriculture will shift poleward. Siberia, for example, which has historically been covered by permafrost (soil that has been frozen for more than two years), is seeing annual temperatures rising over twice as fast as the global average. In the near future, continued climate change may make 1.3 billion hectares of Siberian land available for farming by 2080. In our current global food system, approximately five billion hectares are farmed, so, especially in a less land intensive vegan food system, Siberia could be the agricultural heartland of a climate changed planet.[22] However, if Siberia is allowed to thaw, huge amounts of carbon dioxide and methane stored in the frost and underground would be released, creating an even deeper climate crisis.

Obviously, the future is not going to be this neatly wrapped up. Navigating a transformed food system will still rely on predicting and working within a changed climate. Food systems are already impacted by drought, meaning places like India that are key producers of chickpeas, beans and lentils today are probably not going to be the future producers due to their excessive temperatures. This production might shift to Canada and eventually Siberia, while production in the tropics should halt altogether as these are essential carbon sequestration systems. Future global food system transitions, whatever form they take, will have economic consequences, and will be shaped by geopolitical tensions and create new imbalances. The shifting of food production would not only gain countries like Russia power over significant portions of the world, but would also damage the political position of those who could no longer rely on exports as part of their economy. Global environmental co-operation is already strained, with some nations expressing the hypocrisy of the West dictating how rising economic powers should grow.

While proposed pathways to transition can address practicalities, these models can't consider changing and clashing social, cultural and political influences on food systems. Any scaling and transition to a vegan agriculture, much like any other intervention, would need to rely on careful planning, oversight and political buy-in from states, institutions and, crucially, people themselves. It's crucial to remember that today's food system which seem so natural and unchangeable is unrecognizable to food production one hundred years ago. On a damaged planet that urgently needs intervention, it's no longer unimaginable to propose a *more* vegan world by 2050, but these environmental transitions are not just practical; they need to explore a future that centres values of justice, co-habitation and compassion – for humans, as well as animals and the planet.

* * *

In this chapter, I have looked at how animal agriculture is affecting the planet and considered the alternative vision and impact that veganism has been prioritizing. The continuation of a global food system that relies on suffering, pollution and exploitation is killing life on Earth. While the planet itself will survive, it will no longer be safely habitable by humans or many other species, of which thousands have already become extinct. When thinking about the environment, there is clearly a strong case for transforming our food systems and veganism could play an important role in valuing the natural world and non-human life. It is also worth remembering that while changing our diets might seem radical, the vast majority of food consumed already comes from plants. But despite veganism becoming more mainstream in environmental circles it is still far from the norm, and an environmental ethic alone does not necessarily lead people to veganism.

Environmental veganism has surged at a moment when the future of the planet is uncertain, but the impacts of the human relationship with animals are becoming increasingly undeniable. A growing focus on the impact of global food systems on the planet has seen veganism increasingly in the spotlight, but environmental concerns have long been part of the movement, pre-dating any popular contemporary surge. But, veganism does not just consider the environment as part of its concerns, it is unwaveringly *for* the protection of and building of new relationships with the environment, and the Earth's non-human inhabitants.

4
HEALTH

In his song 'Wha Me Eat', British-Jamaican vegan reggae artist Macka B sings that he is 'Ital' (a natural, organic way of eating often followed by Rastafarians, pronounced as vital, without the 'v') and vegan, avoiding all animal products in his diet, which leads to confusion from people when he talks about his diet. He goes on to list all the foods he eats as a Rastafari Ital vegan, a diet which is intended to increase his levity (the lifeforce of God, or Jah). Consuming foods that are natural, pure and directly from the Earth leads, for Rastafarians, to the nourishment of mind, body and soul, seeking health through food and using food as medicine. In another song, 'Health is Wealth', Macka B tells how this diet changes his relationship with the world to align with the elements, through the compassion shown to animals and the Earth.[1] In his Rastafari Ital veganism, Macka B showcases the promotion of a healthy, high-energy diet with a love and respect for animals, focusing on the body, community and the non-human world at the same time.

The overlaps between veganism and health are often associated with the rise of the plant-based diet that became popular during the 2010s. However, Rastafarianism offers one example of much longer and more diverse histories of vegan eating, having developed in Jamaica in the 1930s as a religious and social movement emphasizing natural and healthy living. This culture of vegan eating and cooking is also present in other religious movements such as sects of Jainism, Hinduism, Buddhism and Judaism, although none of these are as focused on health as Rastafarianism. In addition to cultural and religious beliefs, health studies for at least the last 30 years have found vegan diets to reduce the risks or symptoms of many health problems including heart disease, some cancers and diabetes. Increasing consumption of plants and reducing those of meat and processed foods seems obvious as a way to promote health, but the complete elimination of meat and dairy is far more controversial, with common misconceptions surrounding protein and nutritional sufficiency. Amidst global concerns over diet and health, veganism has seen an influx of interest by those seeking healthier ways to live. In this chapter, I look at how individual health has been part of the growth of veganism, but so has a concern for public and planetary health.

Eating, health and diet

Plants have, throughout history, been the bulk of the human diet, as most people had little choice but to eat what was locally and cheaply available. This has led to diverse local and national cuisines, such as China's development of tofu, now a staple food all over the world, over two thousand years ago; the Japanese-named seitan (a kind of wheat gluten), also dating back to China in the 6th century; fermented soybeans called tempeh from Indonesia; as well as Kellogg's 'Nuttose', a peanut-based meat alternative developed in 1896 to appeal to Seventh Day Adventists. While it is hard to come by exact numbers, a global 2021 survey found that 88 per cent of food industry practitioners expected plant-based food demands to continue to rise,[2] with health concerns being cited as part of people's decisions for becoming vegan. This suggests that the production of novel health foods and their association with veganism spells changes for social understandings of veganism, and also for commercial markets.

Today, because of global supply chains, many vegans can enjoy products from all over the world. The explosion of food imports and exports has allowed for changing and more varied diets, which in turn has expanded what counts as a healthy diet. While malnutrition is still a huge problem worldwide, this is today accompanied by a crisis of over-nutrition, which is the excessive intake of nutrients, which can lead to imbalances and health impacts. The last half-century has seen trends towards diets rich in meats and processed foods, energy- and water-intensive foods, and these dietary changes have been accompanied by new technologies, policies and markets as well as scientific advances in the nutritional sciences.

In a 2013 essay, sociologist of genetics Hannah Landecker wrote a history showing how the science of nutrition and eating has changed over time. In the 19th century, human metabolism was thought of like a factory: food goes in and waste comes out.[3] In the present, post-industrial, era, scientists know metabolism is far more complicated than that, and are more interested in working out how metabolism can be regulated for today's nutritional challenges of both under-nutrition, but also related to an excess of nutrition, which has negative impacts on human bodies and the Earth through extractive food production systems (as discussed in Chapter 3). These transformations in the global food system and in understandings of human health have coincided to produce a whole range of ideas of what a healthy diet looks like, ones that vary culturally, geographically and nutritionally. However, with the linkage of meat and animal products with wealth in many cultures across the world, eating a diet rich in only plants for nutrients and protein, then, challenges many of the cultural and nutritional ideas about what is healthy and desirable to eat in terms of protein, nutrients and calories. This leads to criticism of veganism as being both overly healthy and not healthy at all.

Veganism and health

Veganism has recently seen high-profile attention, from Gwyneth Paltrow's elixirs of youth to Novak Djokovic's plant-based athlete's diet. Veganism, health and wellness now crop up together in more and more different guises, from influencers, people invested in health and exercise and in the form of genuine medical advice. Veganism is widely agreed upon to be healthy at any age, and vegan gym-goers and athletes advocate for its effectiveness in keeping fit, while some beauty and skincare gurus credit a vegan diet as preventing premature ageing. However, health is contentious in veganism, with vegans simultaneously associated with both excessive health and poor health in the popular imagination, as well as harmful health practices.

Health has always been part of vegan lifestyles in one form or another. In the 1960s and 1970s in the United States, vegetarianism and veganism were important to countercultural concerns about the environment and the power of big food producers. Frances Moore Lappé's book *Diet for a Small Planet*, first published in 1971,[4] was an influential touchstone for 'fringe' social groups. In it, Lappé advocates for humankind removing itself from the top of the food chain for environmental reasons, but also speculates on the medical benefits of vegetarian and vegan diets. By the twentieth anniversary of the book, those medical benefits had mounting evidence behind them, with an established correlation between increased meat and processed food consumption in the United States and serious health conditions including heightened risks of heart disease, cancer, diabetes and premature death. Vegan diets tend to be rich in nutrients and low in saturated fats, eliminating carcinogenic animal foods and yielding benefits which researchers are now documenting. Some of the peer-reviewed medical health benefits of a vegan diet include a reduced risk of heart disease, lower cancer risks and a lower risk of Type 2 diabetes.

A focus on health can produce anxiety for vegans about whether they can thrive on a vegan diet. A 2004 psychology study on veganism, health and accountability looked at how an internet forum served as a space for vegans to discuss nutrition, health problems and possible solutions, and resist negative assumptions about health and diet. They focused on how people managed blame and individual responsibility in their food consumption choices and how that shaped their perception of health outcomes. They found that vegans discussed common issues in their diets such as ensuring they were consuming enough essential vitamins and nutrients, as well as seeking support for concerns over health problems. In this forum, 'health' and 'illness' were discussed as opposites and the focus was on solving deficiencies rather than improving health. At the same time as this association between health and veganism, the opposite discourse is also readily circulating: that vegan diets lack protein, calcium, iron and key vitamins like B_{12}, and vegans have long been pre-empting, refining and supplementing their diets to promote health.

Vegans are, generally, cognizant of potential health risks and the importance of a good and varied diet. In fact, a study from 2005 indicates that vegetarians in general are more health conscious than non-vegetarians.[5] This is, however, not to be confused with saying *all* vegan diets are healthy, nor that this is the *most* important part of veganism; the obsession with aesthetics, thinness or muscularity can undermine the radical ethical and political aims of veganism addressed in the previous two chapters. Too great a focus on health and food can lead to uncomfortable places, notably in relation to the idea of veganism producing particular *kinds* of bodies – specifically, thin and White ones. When people go vegan, their families and friends often worry about the impact of their diet on their health – a kind of surveillance that vegans felt was never directed at the healthiness of their diets when they ate animals. This focus on healthiness can therefore undermine the confidence of those transitioning to veganism, especially under increased levels of surveillance on what they eat and how they look.

The focus on health in veganism has been promoted by the rise of vegan influencers who credit veganism for 'improvements' they see in their bodies. A notorious example is that of (now ex-vegan) Timothy Shieff, who rose to small-scale fame in Britain on the BBC's *Ninja Warrior* TV programme in 2015. At the time, Shieff was vocally vegan, regularly on social media and talk shows arguing that veganism was essential to his athletics and muscular body, as well as discussing animal rights and co-founding a vegan clothing company. However, in 2018, following a 35-day period of water-fasting and urine therapy (drinking his own urine), Shieff released a video in which he, crying, admits to no longer being a vegan and having begun eating fish and eggs and wanting to kill animals himself. His conviction in veganism, it turned out, was part of his quest for self-improvement in his health, and many vegans were shocked that a vocal animal rights voice was, after all, uninterested in the ethics of eating animals.

While veganism is sometimes wrapped up with a quest for health, examples like that of Shieff are rare and undermine the ideals that veganism more broadly seeks to uphold. Undoubtedly, there are individuals and groups of vegans whose approach to health is alternative, sometimes unhealthy and even dangerous, but for the vast majority of vegans, health is just one small part of a much larger umbrella of ethics, politics and practices in their lives. This becomes clearer when looking at how ideas of health in veganism are not only about the physical health of humans; there is also a concerted effort to broaden health to include mental health, public health and planetary health.

Feeling good? Mental health

Veganism's impact on health isn't simply about physical health – there are also reportedly mental health and wellbeing effects of becoming vegan. This isn't a recent phenomenon; it has been discussed by vegans for decades. In a film produced by the Vegan Society for *Open Door* (BBC) in 1976, debates around veganism were aired on television at a time when it was rare to be vegan. The film opens on an interview with Dr Frey Ellis who is talking about clinical research with hundreds of vegans. The research found vegans were lighter and had lower blood cholesterol than meat-eaters where both were at normal health. Ellis said the only distinction between vegans and meat-eaters was that vegans 'smile more'. In the 1970s, vegans were considerably more stigmatized and misunderstood than today, not least because there simply wasn't as much visibility.

Across nutritional, health and psychological research into the mental health and wellbeing implications of veganism, the only thing that is clear is that there is no consensus on the health outcomes of veganism. This has not prevented heated debates between vegans, non-vegans and even medical professionals over how far diet can go to treat illness, with mental illness such as depression, anxiety and stress often being up for debate. In 2022, a study in the *Journal of Affective Disorders* undertook a questionnaire with over 14,000 people in Brazil, which found that respondents who didn't eat animals were twice as likely to report depressive symptoms. However, dietician Mary Mosquera-Cochran questioned the legitimacy of the study, which she described as a simple data analysis[6] that had no insight into cause and effect or the reasons behind the results. Vegans lauding their new lifestyle, and especially their diet, as aiding their mental health seem to be as common as vegans and former vegans discussing the toll it has taken on their emotions. When I asked vegans about how veganism made them *feel*, their answers often included physical health but also mental health, both positively and negatively.

Emotional and mental experiences of veganism are not simply individual ones, but also encompass interpersonal relationships and societal acceptance. While veganism has been on the rise across the world, it is still often perceived negatively by the public, the media and medical and health professionals. This bias against veganism has been termed 'vegaphobia' by sociologists Matthew Cole and Karen Morgan, who found that in UK national newspapers in the 2000s vegans were 'stereotyped as faddists, sentimentalists, or in some cases, hostile extremists'.[7] They showed that this bias against veganism led to the misrepresentation and marginalization of vegans, their views and the violent nature of the human relationship with animals. But it isn't just vegaphobia that discriminates against vegans: so do structural biases which can lead to harassment and/or exclusion. It is perhaps unsurprising that vegans who experience discomfort, isolation and hostility in social life might report experiencing worsened mental health.

The association of veganism with symptoms of depression and anxiety is unclear. While anti-vegan stigma might associate mental illness rates with diets, it is more likely that veganism is accompanied by many of the mental health risks associated with activism against injustice. Like many other activists, vegans face multiple pressures from inside the movement, from their social networks and from wider society. It is perhaps no surprise, then, that ex-vegans have consistently reported a desire for social connectedness as one of the key reasons that they have stopped being vegan.[8]

Risky business: public health

When it comes to their physical and mental health, vegans engage in processes of education, research and networking to understand nutrition and improve their wellbeing. The transition to veganism is not simply one of cutting out animal products, but a more wide-ranging transition, which might include exploring health practices such as supplementation of vitamins; expanding and experimenting with food varieties; and learning more about nutrition. This exploration can sometimes lead to more holistic approaches to food and the natural world, such as in Rastafari veganism, but which might also overlap with movements around self-sufficiency, organic eating and the promotion of natural fibres, that have similar sustainable and healthy philosophies. But veganism doesn't just improve individual health, it can also offer alternatives in a world where health is under threat (see Box 4.1).

The impacts of food production on the planet and animals are core concerns of veganism, but these systems of production are increasingly being situated within a broader ecology of violence and harm that is less visible. Concerns over soil degradation,

Box 4.1: More-than-human health

In 1964, American veterinarian Calvin Schwabe introduced the idea of One Medicine to reflect on/promote the importance of thinking about animal and human health together, arguing against the compartmentalization of medicine by species. More recently, this has been taken up in 'One Health' approaches to medicine, emerging in the early 2010s, which advocate and practice medical intervention in ways that see human, animal and environmental health as best addressed through 'harmonized' treatment and prevention.

A model example of a disease that could be eliminated through One Health is rabies, which kills around 59,000 people a year, mostly in rural areas with limited access to education and treatment. A One Health approach focuses not just on treatment, but prevention of rabies through vaccination of dogs, education, investment in health technologies, and developing international partnerships between charities, health organizations, governments, researchers and non-governmental organizations. This plan is known as the Zero by 30 plan to eliminate human deaths from rabies by 2030.

The boundaries between animal species, including humans, are now understood as porous – with zoonotic diseases in particular dominating public health headlines. COVID-19 is the most notorious of these diseases, but avian flu is becoming an increasing concern for epidemiologists. Warmer and wetter climates mean that tropical diseases will move northwards and new ones will emerge in their place.

The One Health approach to medicine is increasingly being seen as a way to address the interconnections between human, animal and ecological health on a changing planet, focusing on technologies to prevent contamination and transmissions between and within species.

Although rooted in veterinarian and medical sciences, One Health also approaches disease as a *social* relation: how diseases spread is related to how and where we interact with one another. This emphasizes the place of education and treatment of both human and non-human species in order to secure a healthy future.

for example, have been raised since the inception of organized British veganism and throughout the 19th century, a burgeoning animal welfare movement across Europe and the United States challenged vivisection and the use of animals in clothing, and in the 20th century, ecofeminists turned their attention to scientific experimentation, the beauty industry and pet keeping. These movements were often intersectional (as discussed in Chapter 7), marrying concerns for animals with wider social justice concerns. What is particularly important when thinking about veganism's relationship with health is that concern has been consistently paid to the risks of animal exploitation and experimentation on planetary health, minoritized people and justice.

Today, the worsening conditions of animal agriculture at the global level are 'creating the perfect breeding ground for diseases to emerge'; agricultural scientist and zoologist Aliza le Roux told *The Independent* that 'our demand for meat is driving cheaper and less controlled agricultural practices, cramming more animals into smaller spaces, feeding them less and less natural fodder'.[9] The COVID-19 pandemic offered first-hand insight into the disaster that zoonotic disease outbreaks can cause at a scale beyond any ever seen before, especially as in the early days of the pandemic, China's wet markets were thought to be the source of the COVID-19 outbreak, similarly to SARS outbreaks in the early 2000s. In September 2020, the United Nations stated that 'COVID-19 is not only a wake-up call, it is a dress rehearsal for the world of challenges to come'.[10] In the aftermath of a global pandemic, greater attention has started to be paid to the significant risk of virus mutation and transmission in the factory farms that house agricultural animals.

Epidemiologists explicitly point to the intensification of animal agriculture as a driver of risk for pandemics. Most notably, highly pathogenic avian influenza (HPAI, or bird flu) 'typically emerges in commercial poultry farms by conversion of an innocuous wild-type virus – aptly called low pathogenic avian influenza virus – into one that causes high mortality'.[11] Science writer David Quammen writes in the *New York Times* that the passing of a mutated version of the virus to wild bird populations has produced shocking and disturbing images of dead birds, birds falling from the sky and encounters with sick birds.[12] A direct result of bird flu mutations in chicken and turkey factories, this disease is now killing a range of species, such as bald eagles, condors, gulls, terns, hawks and falcons, many of which are already endangered or under threat from climate change. Anthropologist Frédéric Keck explores in his book *Avian Reservoirs*[13] the threat of bird flu viruses crossing to human populations, which is a serious concern and, in the summer of 2023, there were at least four cases of transmission to humans.

Contemporary concerns over zoonotic interspecies risks are not new, with high-profile pandemics of swine flu in 2009 and bovine spongiform encephalopathy (BSE, or mad cow disease) and its human variant, Creutzfeldt-Jakob disease, in the 1980s and 1990s. However, today's novel viruses are situated within a changing epidemiological context of antibiotic resistance. The concern for public health, then, is not only about the increasing risks of emerging zoonotic diseases, but also the ability to treat them, in human populations. While cases of bird flu in humans are starting to be identified, these are mercifully rare and have not yet been shown to transmit between humans. In wild bird populations, the 2020s have seen devastating losses to bird flu, but it is within animal agriculture that the numbers of casualties are most shocking – and hidden. When a HPAI virus is identified, there are strict practices followed to treat the infected area, which means that 'death is almost guaranteed within 48 hours' for any birds in the area.[14] With some chicken farms holding hundreds of thousands of birds, any outbreak can spread swiftly and see catastrophic loss of life. In the 2021–2022 outbreak, 97 million birds, mostly chickens, were killed globally (with

50 million birds killed in Europe), while the 2022–2023 outbreak, the worst yet, saw 15 million domestic birds dying of HPAI with a further 193 million birds being culled. It is also worth noting that in the human cases of bird flu, these are often in poultry workers. From the perspective of vegans, this loss of life is a tragedy already occurring, not one of a potential future risk. As journalist and vegan activist Claire Hamlett has written, public and media reporting on avian influenza consistently 'fails to mention the likely role played by intensive farming ... in both making the virus more deadly and passing it back to wild birds'.[15]

The interest in and advocacy for greater attention to the structures and exploitation of animals underpinning zoonotic and broader health risks is increasingly important to vegan advocacy and attention. In a 2022 manifesto on veganism and public health, scholar Niñoval Flores Pacaol[16] argues that public health concerns should not and indeed cannot be limited to humans through any empirical or logical position, and that the interactions between public and ecological health are increasingly urgent as debates around planetary futures heat up. Crucially, health is no longer able to be treated in a vacuum as knowledge around the complexities of interactions between bodies, ecologies and wider environments are understood, biologically, metabolically and philosophically. This expansion and transformation of knowledge about individual and public health is contextualized within a set of interlocking and impending crises, of pandemics, nutrition, but also the aggravation of the climate crisis.

The expansion of concern through veganism broadens and extends health beyond the human, into *more-than-human* planetary perspectives.

With the climate crisis having increasingly obvious impacts, people are seeking out new ways of protecting and acting to conserve human and non-human futures. This focus on the environment is not divorced from the health-based goals of veganism but is very much connected. It is impossible to separate the environment from animals or public health as they are all interconnected. The question of planetary health has become a topic of debate for medics and scientists, with an article in medical journal *The Lancet* in 2014 calling for action in relation to threats to health, sustainability and the natural world to be considered under a 'movement for planetary health'.[17]

* * *

Thinking about health as individual, collective and planetary aligns closely with the goals and vision of veganism, and it has been argued by nutritional scientists that plant-based diets offer a promising way of alleviating health and environmental burdens associated with current global diets.[18] Rethinking human diets requires mass transformation, and there is a growing consensus that a plant-based diet offers a possible solution to the diet–environment–health trilemma. Despite meat-eating being consistently linked with poor human, environmental and animal health, there is still huge resistance at individual and policy levels to think about plant-based dietary transitions, and considering veganism as a holistic practice meets even more hostility.

With human, non-human and planetary health firmly on the agenda, it's clear that veganism still has much to offer in a variety of contemporary conversations and action for the future. In this chapter, I have shown why we need to think about health veganism beyond the individual and explored the interdependencies between physical, mental *and* public health. In so doing, what health veganism is *for* shifts beyond an improvement of the self and instead projects visions of collective health and wellbeing – for humans, animals and the planet.

5
CULTURE

Veganism explores how the innovative and playful use of plants in human diets can produce exciting new foods and meals. From bean burgers to cashew cheese, vegan chefs and home cooks have been pioneers in exploring the wide range of uses and forms that plants can be transformed into. As veganism has taken off, there has also been a surge in new kinds of products in the market. From vegan steaks that bleed to replicate animal flesh, to vegan ice cream made from real cow-milk proteins without the cow, lab-based food production is touted by some commentators as holding the possibility of a vegan planet, without people having to give up on the textures and flavours they love. When policy makers and scientists write about sustainable food futures, they are often primarily focused on technologies, land use and agricultural practices that will produce enough food in a climate-changing world, measured in calories and nutrients.

I talk about food technology in the next chapter but, before doing so, I want to emphasize in this chapter how, as humans, we don't eat only to fuel our bodies. We eat to connect, to socialize, to comfort and to enjoy. Food is wrapped up with our social, political and cultural lives and identities, and the consumption of animals is often a feature of social food consumption: just think of turkeys at Christmas or Thanksgiving, or beef burgers on barbecues in the sun. The idea of dietary change can therefore be perceived as a threat to social and cultural norms, which is one reason that vegans might experience hostility from their loved ones, because their changing diet represents a changing identity, which can be jarring and confusing, and food is such an integral part of our emotional and social lives. The focus in this chapter is therefore on *culture*: how veganism is disruptive to meat-centric cultures but can also foster more sustainable and equitable futures.

Food and culture

Veganism changes relationships to the world and other people, both practically in what we eat, but also emotionally and socially. Being vegan can be inconvenient. There is a great deal of work involved in learning what is OK to eat, finding hidden animal products in fortified foods, ensuring access to food when travelling, but also in explaining and defending the ethics and practices of veganism. Vegans are therefore dealing with multiple and competing challenges as they navigate the world: distress at the suffering of animals, discrimination in society and a constant need to defend themselves against critics. At the same time, they experience changes in their own relationships and traditions as they navigate these ethics and practices. Food is undoubtedly the most visible and fraught space where vegans see the greatest transformation, because we all have to eat and, often, we do so with other people. Becoming vegan isn't just a refusal to consume animal products; it also requires the navigation of society and culture in a changed way, which can lead to both challenges and opportunities for advocacy.

Culture is integral to understanding veganism – and this can be seen most clearly when looking at food, but it's relevant to other kinds of vegan consumption too. The potential of plant-based food systems to alleviate environmental stressors and combat climate damage, particularly through reducing carbon emissions (as discussed in Chapter 3), has contributed considerably to the increasing visibility of vegan food consumption and debates around dietary transitions. But food (and all kinds of consumption) is about more than nutrition, and it can show how veganism can transform relationships with culture. Understanding how veganism can become culturally acceptable is going to be key to creating a central role for veganism in the future. It is crucial, therefore, to understand the complex systems in which food is produced not just as material products, but as culture.

Food is intimately tied up with, and produces, culture. While discussions of the importance of food often lead with the importance of health and sustainability, the role of history and culture also shape what we eat, and therefore what future food systems should value. Food can provide entryways into conversation and debate on almost any issue, not least because of its inescapability, essentiality, to human and non-human life. The philosopher Annemarie Mol has argued that eating is usually treated as a lesser activity than thinking, but it shouldn't be, because it can connect us to history, emotion and politics. The kind of foods we eat, how we prepare them, when we consume them, even our taste preferences are shaped by our environments and heritage: the smells, sounds, textures, appearances, marketing, origin, timing and placement of foods affect how we feel about them. Most of us probably wouldn't regularly enjoy a pizza at 7am, but a cold, leftover slice when waking up with a hangover might be just the antidote. A roast dinner may not be your favourite meal, but when enjoyed with loved ones on a festive day, it might be the best thing you've ever tasted. Eating a meal, whether shared or alone, is always informed by our wider lives, cultures and relationships.

Food and taste are cultural but, for vegans, navigating these food practices often means they are in contention with dominant cultural norms. The relationship between societies and what they eat has been long understood by critical food scholars as an evolving one. For vegans, the consumption of animals as food is connected to other forms of hierarchy and oppression. Carol J. Adams's work on *The Sexual Politics of Meat*,[1] for example, has shown how patriarchal societies sexualize and dehumanize women and animals in similar ways, with media, advertising and politics degrading women and animals (discussed further in Chapter 7). As well as comparisons between women and meat, there are also cultural associations between masculinity and eating meat, from gaucho culture in Argentina and Uruguay to Vietnamese dog-eating culture, hegemonic masculinity and meat-eating have a long transnational relationship. Veganism presents a challenge to cultural norms around eating, and in the case of gender, can be subversive. When it comes to food, it isn't just about what it means to refuse to participate in eating animals, but about developing new cultures around food.

Vegan food cultures

In the popular imagination, vegan food is paradoxical. It is thought of as simultaneously both healthy and unhealthy, overly concerned with nutrition but also failing to provide essential vitamins, and as being over-processed while also restrictively expensive due to its reliance on wholefoods. Vegan food has not, until very recently, been commonplace anywhere outside of vegetarian and vegan spaces, which has led to it being dismissed as both inaccessible and basic. Vegan food, much like vegans, challenges cultural norms around eating by being an 'alternative'. This is despite, as I mentioned in Chapter 4, foods such as seitan, tofu and tempeh having histories dating back centuries. These foods aren't *alternative* proteins, but a staple part of diets in the greater China region, with soybean domestication and cultivation widely believed to have taken place in East Asia 6,000–9,000 years ago. It is only their importation and growth in the West that has reclassified these traditional foods as *alternatives*, not just creating a sense of otherness of the foods themselves, but also marginalizing the communities and geographies they come from to the dominant White, Western norms of eating – and culture.

The ethics and motivations of veganism can dominate debates, but with food being such an intimate and important part of life, there have been emerging voices and experimentations with showcasing the potential of vegan foods not as 'alternative', but as appetizing and creative in their own right. Food writers such as vegetarian Alicia Kennedy have advocated for not just ethical ways of eating and connection to what we eat, but for food as both lifestyle and ideology. Similarly, taste and flavour are integral to the work of chefs trying to make vegan food mainstream, dispelling myths around veganism as tasteless. Puerto Rican chef Lourdes Marquez-Nau converted the menu of her restaurant Casa Borinqueña in California to a totally vegan one in 2022, having become vegan herself during the pandemic. Speaking about her approach to vegan food, she said to San Francisco radio station KQED-FM, 'meat doesn't have a flavor. We introduce the flavor. ... It's a matter of having the same approach [when cooking vegan food]'.[2] She infuses vegan meats, such as soy-based patties, with traditional seasonings and cooking methods, making veganism work with other forms of cultural taste. After recently closing their main location, in 2024, Casa Borinqueña opened in the San Francisco IKEA food hall, bringing vegan food to different people.

The association of veganism with a particular kind of lifestyle, notably one around health, and therefore pleasureless foods, is also one that is being challenged by vegan chefs. In a 2023 profile in *The New Yorker*,[3] founder and CEO of Atlanta's Slutty Vegan restaurant, Pinky Cole, talked about her vision for the restaurant to be food with personality, a 'taboo experience', born out of a movement of Black veganism but now trying to reach everyone with their indulgent food, full of pleasure – and sluttiness. The marketing of Slutty Vegan brings a different kind of vibe to perceptions of veganism as a deprivation of joy. The burger chain sells plant-based patties from Impossible Foods, a plant-based food company famous for its use of heme, a molecule that makes blood red and makes the company's burgers 'bleed'. Its financiers include Bill Gates and Google and, in 2019, it partnered with Burger King to release the Impossible Whopper. Ethically problematic, perhaps, but undeniably fun, accessible and showing how veganism isn't antithetical to joy – and cultural diversity.

At the same time as fast, unhealthy food is sweeping across the vegan food scene, higher-end restaurants are also focused on bringing vegan cuisine onto their menus and expanding their customer bases. The three Michelin stars received by New York's Eleven Madison certainly don't come from serving up the same kinds of food as Slutty Vegan or Burger King. Rather, Eleven Madison is celebrated as having 'bold visions of luxury dining' that bring experimental and gastronomic ideas together in their plant-centric menu. Meat alternatives like the soy-based patty tend to be absent from this kind of high-class experience, with vegetables the star of the show. In Manchester, United Kingdom, the Allotment Vegan Eatery has a similar vision, to source from local harvests and create vegan food 'like you've never tasted'. This showcasing of vegetables, as opposed to the somehow obscene fake meat, is also notable on the vegan menus of expensive omnivore fine dining restaurants. The upside of this is a broader acceptance that vegetables can be the main event, not just a side dish, with unique and innovative techniques of serving them. However, it also plays into stereotypes of veganism as exclusive and expensive, a dining experience for the well-off, while the masses get bean slop and rice.

The diversifying opportunities for vegan food consumption speak to the growth of veganism and, despite continuing stigma in society of veganism, a potential hope that in the future, veganism might no longer be an alternative culture but embedded within or at least featuring on the cultural landscape of eating, from the dirty to the indulgent. Veganism is becoming socially, culturally and economically accessible, but the emotional experience of eating as a vegan has received less critical attention.

Emotions and culture

Key to distinguishing veganism from other diets with an ethical impulse – from vegetarianism, and from plant-based eating (which might look the same in practice) – is its vision for a different world. Not just a healthier, or more ethical one, but one that changes the relationships we have with one another, with animals and with the natural world. Veganism is not simply about what we eat, but rather about how we *live* – with ourselves, with each other, and with the non-human world. There is therefore a strong emphasis on situating what we eat as part of who we are: veganism is essential to our identities and is expressed most strongly in what we eat (and don't). But what we eat isn't just experienced practically, it is also an *emotional* activity.

While there are many pathways to veganism, some taking a long time and some happening instantly, what all of them share are moments of revelation on those journeys. In a research paper from 2000, Barbara McDonald explores veganism as a change made when people can no longer 'unknow' knowledge about the violence that humans inflict on animals, which she argues follows a process that includes 'catalytic experiences, possible repression of information, an orientation to learn, the decision, learning about veganism, and acquiring a vegan world view'.[4] One vegan who I interviewed during my research exemplified this process; she had begun changing her diet on the advice of her doctor and, after cutting out many different foods, was practically following a vegan diet. On her journey to health, she began researching vegan diets and, unintentionally at first, stumbled into online spaces and forums that taught her more about how animals were treated in farms and other industries. She begun actively seeking out more information, despite finding it deeply disturbing, with disruptive impacts on her everyday life:

> [W]hen I first had that revelation moment, it was quite difficult. I cried at random stuff. I was reading and reading and reading. I was listening to 'Farmageddon' when I was driving to work, and I'd have to pull over and be like 'oh my god the bees! The bees!' as fresh waves of realization and horror came over me.

Figure 5.1: Drone footage on factory farms, 2015

Farmageddon: The True Cost of Eating Meat is a 2014 book by Philip Lymbery (the CEO of Compassion in World Farming) and journalist Isabel Oakeshott that looks at industrial mega-farms across the world, such as this one photographed by Anima International in 2015.

The growth of contemporary veganism cannot be understood as a justice movement that is straightforwardly rooted in rationality, as the 1970s animal rights movement contended (and as I discussed in Chapter 2). There is a strong role for emotional politics in veganism today, one that focuses on veganism as visceral, rooted in gut feelings of right and wrong. Food has the power to invoke emotions, collectively and individually, and for vegans, animal products can invoke particularly strong reactions of disgust, anger and upset. Feminist thinker and writer Sara Ahmed argued in her book *The Cultural Politics of Emotion*[5] that emotions can lead to collective politics and social alliances, and veganism is certainly full of emotions – whether of anger, sadness, rage or joy and hope.

The embodiment of veganism, discussed in Chapter 4 in relation to health, also manifests in reactions to food, both physical and emotional, particularly where things like disgust are experienced both mentally and in our bodies. Veganism produces a new relationship with the world, which can be felt most keenly in relationships and encounters with food, because food is so central to everyday life. While, over time, these reactions to the suffering of animals might become less emotionally shocking, they still persist and contribute to psychological distress for many vegans.[6] This can be seen clearly in the reactions of vegans who have accidentally eaten animal products.

In August 2023, vegan singers and sisters Chloe and Halle Bailey accidentally ate meat for the first time in over a decade of veganism after they were mistakenly served beef instead of vegan Beyond Burgers (a beef replica patty). On a TikTok recorded by Chloe, she recalls the juice dripping from her one bite of the beef burger and, after immediately calling to check if they were vegan, and finding out they weren't, she starts crying and 'freaking out'.[7] This reaction may seem extreme, but the consumption (even accidental) of animals comes with a huge amount of guilt and revulsion for vegans, and this is a common experience shared on online vegan groups, particularly on Facebook. There are hundreds of vegan Facebook groups for places across the world and different kinds of vegans but, of those I am a member of, pictures being posted of food to ask whether others think it is definitely vegan happens many times a day (particularly when the food comes from non-vegan chain restaurants) and, almost as frequently, there are posts of people who are sick and horrified at accidentally eating animal products.

Experiencing guilt, shame and disgust at accidental consumption of animal foods is, obviously, not healthy, but it is also symptomatic of clashing social and cultural values. Vegans find it hard to exist in societies and cultures that don't seem to understand them. These exclusions from mainstream society, particularly before veganism became more common, led to the blossoming of local vegan communities. Many universities and colleges have active vegan societies, for example, while groups like Vegans in Taiwan, Vegans in Warsaw, and many other places around the world welcome both tourists and locals to potlucks and meet-ups at vegan eateries. The experience of being vegan, then, can be both isolating *and* an opportunity for connection, both online and in the 'real' world, where the highs and lows of veganism can be shared in a safe and understanding space with others who *just get it*.

Nature, culture and veganism

For many people across the world, food isn't easily available or abundant. The globalization of food systems over the last century has led to people becoming more disconnected from the production of food (see Box 5.1), exemplified in the majority of food now being produced by large agribusinesses. The pace of life has also changed over the last few decades, with the acceleration of technology, work and leisure meaning that convenience food has become a greater part of the human diet. Scientific advances in nutrition have seen foods enriched with vitamins and minerals, and natural, locally grown foods are a long way from the diets of many people. Today, the human population is simultaneously facing crises of under-nutrition and starvation, and over-nutrition, often in the same countries. And all of this is happening amidst global environmental crises that are putting fragile food systems under increasing pressure. Yet, even within this fast-paced convenience lifestyle, food can be a source of connection and comfort: the taste of home, of childhood; a step into another culture; solace or exploration.

Box 5.1: Veganism and Indigenous practices

Veganism is sometimes criticized for being hostile to Indigenous ways of life that rely on subsistence hunting and fishing. Critiques of veganism as anti-Indigenous point to animal rights activism targeting the seal-skin trade by Inuit communities or of small-scale hunting in North America. These arguments are often deployed as a reason that veganism is violent and colonial.

However, looking at the history of land dispossession and deforestation of once-Indigenous lands, this is far more commonly connected to land clearance programmes for the grazing of cows or for growing soy to feed animals in industrial agriculture. Contemporary food systems and access to food have disproportionately negatively impacted Indigenous communities. For example, the Navajo Nation is a food desert with few grocery stores and little access to fresh vegetables. Processed foods, especially meats, therefore, make up a large proportion of the diets here, albeit supplemented with hunted food. This has led to five times the prevalence of heart disease than in the general population of the United States.

Colonialism broke relationships between humans, animals and the Earth and created/replaced them with extractive relationships. There is no excuse or place for anti-Indigenous sentiments or practice within a veganism that promotes justice for humans, animals and the planet. In fact, the relationships that Indigenous communities have with the non-human world are ones that many vegans respect and aspire to: compassion for animals, and caring relationships with the environment.

Indigenous vegan activists, such as Genesis Butler, are increasingly advocating for the synergies between vegan eating and Indigenous values. Jen Rivera Bell shares her reimagined plant-based traditional foods on her Instagram to show how food culture can persist through veganism. Professor Maneesha Deckha has addressed this false opposition of veganism and Indigenous interests by showing how veganism can reject projects of settler colonialism such as animal agriculture and be a way of decolonizing diets.

While there may be tensions between Indigenous hunting practices and veganism, there is also the potential for constructive engagements between the two over their shared commitments and visions of the future.

Veganism isn't a singular food culture, and nor is animal-eating. However, it's important to address ideas that veganism by definition breeds a special or more connected relationship with food and, more broadly, nature. While almost every vegan will have changed what they eat, it doesn't necessarily establish a new set of values. Setting veganism up as in opposition to mainstream culture is a false dichotomy, even though it may be true for some vegans. Especially with the rise of processed vegan meats, dairy and cheeses, a vegan diet may not look all that different from an animal-based one (albeit without the animals). Today, many vegans can easily buy burgers, sausages and faux-chicken pieces, but also life-like prawns, salmon and scrambled egg. Some of these products have been designed to reproduce the fatty, bleeding and streaky nature of actual animal meat. It could be argued that there isn't a particular investment of the consumption of these kinds of foods in rethinking our relationship with nature through food, although they are still better for the environment than animal products and are widely consumed by non-vegans as well. There are endless other kinds of food culture and relationships with animals that both do and don't rely on eating them, understanding food *as* cultural and social as well as nutritional is key to unlocking sustainable future food systems.

* * *

Within this complex cultural landscape, veganism is both an opportunity and a threat. The growing presence of veganism across the world is leading to backlash as it challenges the human relationships with non-humans, and also with each other through the radical idea that animals and humans shouldn't be harmed in the production and consumption of food. Veganism therefore disrupts cultural norms and rules through pointing out the harm and suffering underpinning many traditions, particularly in our food consumption.

The threat of veganism as a cultural practice is also political: veganism challenges large-scale food systems in which corporate and state actors have huge amounts of investment and power. The cultural and political dangers of veganism, then, are both interpersonal in feelings of rejection, and also structural. For example, in the United States, the meat industry is a centralized political force that has influence over laws and regulations.[8] While there are some famous vegan politicians in the United States – namely Cory Booker, the Senator of New Jersey and 2019 presidential candidate, and the former governor of California and bodybuilder Arnold Schwarzenegger – they both leave animal welfare firmly outside their political roles. Booker especially has been criticized for not advocating more strongly in his debates for veganism and against animal factory farming, despite having a strong plan for animal welfare. His reluctance to be vocally pro-vegan might be related to the political and financial power that Big Meat exerts in US politics.[9]

Hostility towards vegans at the individual, social and structural levels is often based on misunderstandings or stereotyping of veganism, within a context of 'alternative' ways of living often being treated with suspicion or meeting ridicule. Veganism changes relationships with food and although it isn't the only thing to do so, its growth and potential is multi-scalar, both at food system and individual level, and must be understood as cultural as well as practical. In the contemporary moment, it is abundantly clear that the food that we produce and consume is unsustainable. Many different visions of future food systems and relationships exist. Some advocate for a return to smaller or regenerative forms of farming, while others (as discussed in Chapter 6) build on technological hopes for a revolution in how we eat. This chapter has looked at why food as culture matters, and how this is particularly important in veganism, even though veganism isn't just about food. Veganism, then, is not just about food practices, but is *for* cultural, sustainable and ethical modes of living and experiencing consumption not just as individuals, but as a collective.

6

TECHNO-VEGANISM

With the increased visibility of veganism and a growing plant-based foods market sector has come an opportunity for technological innovation in alternative proteins. A report by researchers at the Washington DC-based think tank, the Center for Strategic and International Studies, states that the label 'alternative proteins' can apply to a vast range of foods, 'from plant-based protein, to cultivated meat grown from cell cultures, and even to insects'.[1] On the market today are a range of burgers, minces, chicken, sausages, cheeses and eggs, all made from plants. And on the horizon are products made from cellular agriculture: meat and dairy without the killing of animals. The alternative protein market covers both plant-based and food-technology-created non-animal-based proteins, but there are differences between the two (see Box 6.1). The contemporary interest in the consumption of alternative proteins stems from concerns over the impact of food production, but

Box 6.1: A menu of fake meat

Plant-based proteins are made from grains, legumes, nuts, fungus and algae. An example of a plant-based protein is Quorn, a novel mycoprotein made from fungus that has been on sale since 1985. Tofu, tempeh, seitan and soya-based products also fall into this category. These are foods with specific cultural and geographical histories. However, with growing consumer interest in non-animal proteins (from vegans, but also from wider ethical consumer markets), there has been growing investment in technologies that seek to replicate the experience, texture and taste of meat.

Beyond Meat, for example, was founded in 2009 and launched its first products in 2012, as a meat alternative that seeks to replicate the qualities of animal flesh through a novel combination of soy, wheat and carrot in its chicken strips, and pea, mung bean, rice, canola, potato and pomegranate in its burger – notable for being the first vegan product on the market to 'bleed' through the addition of beet juice. These plant-based proteins are readily available in markets across the world, both the traditional and newer iterations, with novel developments producing plants that get ever-closer to the real thing.

Food-technology products are not, however, made from plants, although many believe they will lead to an uptake in veganism when they reach the market, as they do not come from animals. These kinds of meat – sometimes called cellular meat or cultured meat, are quite literally meat without the animals (in the final product, at least – donor animals are still needed to provide cells). They are produced by culturing animal cells in vitro and rely on novel cellular agriculture techniques that produce meat from cells in a bioreactor, rather than animals. These in vitro cells are fed and grow into animal flesh, without any sentience and therefore no ability to feel pain.

In 2023, Australian company Vow took this even further and produced a meatball made from the long-extinct woolly mammoth by taking the DNA sequence for mammoth myoglobin and combining gaps with elephant DNA. Vow CEO and co-founders said that the mammoth meatball was produced as a symbol of biodiversity loss and a call for the need to change our diets.

While *plant-based protein products* have been commonly consumed for decades or even centuries, there is far more apprehension aimed at the future of novel food technologies that will more accurately replicate meat.

In the European Union in 2020 there was even pressure to ban vegetarian products from being sold as 'burgers' or 'sausages', despite these referring to the *shape* not the *ingredients* of a given product. Although the ban didn't pass, there are rumblings of dissatisfaction and fear, especially from animal agribusinesses, who are concerned that increasingly ethically conscious consumers might choose meat without the murder.

also, its vulnerability. Within a global context of uncertainty and crisis across health, the environment and geopolitics, the global food system is constantly at risk of collapse, and it is within this context that alternative proteins have swept onto the stage with environmental and security promises. In this chapter, I will be discussing what I call 'techno-veganism': the coming together of vegan ethics and practices with scientific innovation.

'Alternative' proteins

Techno-veganism promises innovation and expansion for the food sector. While plants have provided an important – and often innovative – source of protein, there continues to be a perception of diets without animal products as insubstantial. This association of plants with low protein is a gendered one that is used to emasculate male vegans, for whom a muscular, high-protein body is depicted in the media and on social media as the ideal body. The notion that plants are nutritionally deficient is contradicted by the heavy market presence of plant-based protein powders, such as whey and pea proteins. The idea that humans should be consuming a high-protein diet is a cultural one, with high-protein diets being popular on and off since the 1960s for weight loss, notably in the Atkins diet of the 1990s. The association of veganism with nutritional, and therefore muscular, deficiencies is a stereotype that was taken on by vegan athletes and celebrities in the 2019 film *Game Changers*, whose central premise was that veganism wasn't just healthy, it was a hack to elite fitness. This focus on muscularity and health by way of protein is not confined to veganism. Protein consumption has seen rapid increases in some markets, with regional variations such as high pork consumption in China, dairy and eggs in the Middle East, and those in the United States continuing to eat twice as much beef as the global average.[2]

In 2014, two vegan engineers began experimenting with how milk proteins could be produced without animals and, funded by private investors, founded a company called Perfect Day to change the process of dairy production, without altering its taste. Using precise DNA sequencing of microflora, which is then fermented with water, nutrients and sugar, the company can produce cow whey protein, without the cow. Perfect Day has partnered with a host of companies to bring its precision fermentation to the market, from ethics- and environmentally-focused companies like Brave Robot, which makes ice cream, to multinational corporations such as Nestlé. These novel technologies are not, then, imitating animal products, but literally recreating them without the animal. Proponents argue that the production of food in laboratories frees up land for wildlife and could also produce food systems that are less vulnerable to environmental change than our current ones.

These new technologies go beyond food with applications for products in, for example, plant-based leather made from pineapple, mushroom and cork. The same cellular technologies being used for cultured meat can also be applied to substances such as wood, opening a whole range of possibilities for materials to be produced without the destruction of the planet and its nature, depending on sustainable sourcing and processes. However, as pro-'labriculture' vegan scholars Jan Dutkiewicz and Gabriel Rosenberg[3] argue, people love eating meat, and cellular agriculture offers some hope, some concern, but nothing 'predestined' about the future of food. In fact, they warn against viewing cellular agriculture as a 'silver bullet' and the risks of both techno-optimism and technophobia in the future of food systems.

Techno-veganism is underpinned by a focus on efficiency; the unquestionability of modernization and growth; and the intention to turn a crisis into an opportunity. Critics of new forms of cellular agriculture astutely point out, therefore, that switching to a reliance on technologically produced food rather than animal food does little to shift exploitative relationships with the non-human world or challenge the corporate dominance of contemporary food systems. This critique is grounded within a broader political and scientific context that has relied upon technology as a saviour and ignored the role that policy, social and cultural change has to play in creating sustainable futures.

Technology and the climate

Interventions addressing the impacts and consequences of climate change have long been critiqued for an over-reliance on technological solutions. The political scientist Martin Jänicke has conceptualized the move to technology-focused solutions to environmental issues as modes of 'ecological modernization', which reframe problems as challenges to be solved through human innovation and markets, not a disastrous consequence of human action. Ecological modernization looks at things like clean energy, elimination of pollution and production as in need of technological fixing to reduce their impact. It has also become increasingly common to turn to technology when thinking about how to undertake agriculture in the climate crisis. As we saw in Chapter 3, agriculture, particularly industrial animal agriculture, has had catastrophic impacts on the environment and the climate. But it is also big business, and crisis provides an opportunity for innovation and growing markets, and for making a lot of money.

The aim to remake the planet through deliberate manipulation of environmental processes is known as climate engineering. These ideas and technologies are almost always controversial, with plans like shooting diamonds into the sky, giant space mirrors and iron filings being pumped into the ocean being met with disbelief, despite being *potentially* viable. This is not least because once we start meddling with the climate, it would be impossible to stop and there could be any number of disastrous unintended consequences.

In the climate crisis, the carbon impacts of the methane emitted from beef and dairy cows have been the target of these engineering practices. Animal farming is responsible for around 14 per cent of human-induced climate emissions,[4] and cow farming makes up a big chunk of this, so the potential for impactful interventions is big – and potentially lucrative. While vegans would advocate for the eradication of the animal farming sector, the ecological modernization approach dictates that business as usual should continue, albeit with significant tweaks that are palatable to states and big business. This has led to a plethora of climate engineering interventions that aim to alter the biology and metabolism of cows so that they produce less methane through more environmentally friendly burps and farts, and can continue being farmed and eaten, without the guilt. Scientists have experimented with everything from dietary supplements of seaweed to gene-editing technology, with a face mask for cows that would neutralize burps winning the support of the Terra Carta prize in 2022, an award launched by King (then Prince) Charles in 2021 to recognize climate innovation.

Veganism's ethics and politics are in obvious conflict with these applications of climate engineering of animals – not least because there is usually an inverse relationship between the measures of environmentally friendly farming and the health and wellbeing of animals. But the promise of technology has not escaped veganism, not least because the rise of alternative meat products is challenging the animal agriculture industry and offering, for some, a source of hope in interventions to vulnerable food systems.

Capitalism and critiques

Although, undoubtedly, the wide-scale uptake of cellular agriculture would prevent the deaths of millions, if not billions, of animals, there are still significant technical concerns that complicate what might, on the surface, appear to be a positive trajectory. Perhaps the largest issue is that some of these technologies are not actually in use, but remain speculative, and therefore assessing their potential in food futures is difficult (see Figure 6.1). For vegans, the technical issue of the continued need for donor animals for bioreactor production of meat is the predominant concern, but there are also issues of scalability for these novel practices. The sourcing of cells, mimicking of the womb environment in laboratories, and the processing of tech-meats at commercial-scale are the key scientific challenges. Even if these are addressed, there remain political and institutional hurdles for the sector, as regulatory systems do not yet exist.[5] So, while consumers may be keen to get their hands on some cell-based meats, there are still complex technology and policy paths to navigate. There is also the question of who stands to profit – and whether they have the political and economic power to successfully overcome these challenges.

Food is big business. Investors see the challenge of feeding a growing human population and dwindling natural resources not as a space of crisis, but one of opportunity. It's therefore not surprising that one of the biggest critiques of these technologies is, in fact, that they don't necessarily envision a radically different food system to the one that we have today,

Figure 6.1: Cultivated meat production process

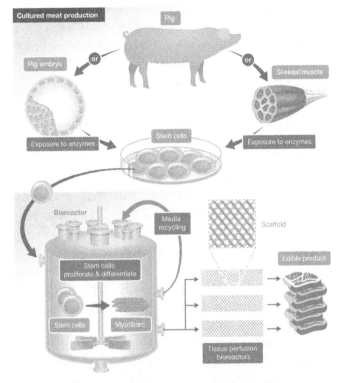

Processes of making meat in labs are complicated and expensive. They rely on technology like huge bioreactors to grow cells as well as scientific knowledge about genetics, metabolism, cells and chemical reactions and industrial resources to scale up and process. This image, produced by Tuomisto (2018) shows how cultured meat takes stem cells and differentiates them into muscle cells, before putting them into a bioreactor to grow in number. These cells are then put onto a scaffold to grow into meat products.

not least because access to these forms of production will be prohibitively expensive. The concern is that technological patents and power will centralize in the hands of the same conglomerates and corporations, driven by market metaphors as 'biotechnologically oriented economic activity ... as the key to solving global food systems changes'.[6] Leading CEOs in the alt-meat business don't see themselves just as technological innovators or entrepreneurs, but as political activists, while multinational food corporations like Tyson Foods (the world's second-largest processor of chicken, beef and pork) are 'willing to participate in [their] own disruption'[7] to get in early on this potential new site of profit. This has led to concerns not only over the nutritional and safety considerations of tech-meat, but also over the control of food systems – and how they might consolidate power rather than disrupt it, despite the calls for radicality embedded into the food technology sector.

Cellular meat, even if it is possible technologically, may not be economically viable without these kinds of corporate investment, due to the lower quantities of food currently able to be produced for the same cost as cheap meat, which has been subsidized and intensified over the last century to artificially keep prices low. In 2021, a techno-economic analysis run by Dutch independent researchers projected that, dependent on technical and economic barriers, the production price of cellular meat could drop from $10,000 per pound (or 0.45kg) today to $2.50 per pound by 2030.[8] However, critics have dismissed the analysis as unrealistic and missing key parts of the production process, underestimating costs and misrepresenting the complexity of scalable facilities. This is something cellular meat producers and advocates like the Good Food Institute are, to their credit, honest about when discussing the challenges of technology and economics, setting an expectation that investors should be prepared for a smaller profit than traditional meat – which has been modified, bred and intensified for over a century to extract value at the expense of animals, nature and human workers. Despite these warnings, there has been a total of $2.8 billion invested in cultivated meat and seafood companies, with $896 million of that being invested in 2022, across 156 companies.[9]

Novel food technologies are perceived by investors to have huge financial potential, and this has rightly led to concerns over a 'race to market' that focuses on being the first to launch a product and capitalize on this anticipation. Dwayne Holmes, of a Sacramento-based donor-funded research institute into cellular agriculture, has suggested that the race shouldn't be to market, but to *mission*. Holmes and colleagues across academia and the private sector highlight five parameters of this mission:

1. slaughter-free and transparent production processes;
2. stringent regulatory processes to prove safety;
3. address sustainability concerns;
4. be provably able to scale and price for sale; and
5. practise open science that refuses patents and mobilizes researchers.[10]

This vision of collective and robust techno-ethical production is not beyond the realms of possibility, but the chances of lab meat being greenwashed rather than a force for genuine ecological restoration is a real concern, particularly in a context where food – and threats to food systems – is not just a nutritional issue, but a political one.

The politics of eating

Techno-veganism advocates believe that animal products can be produced in a way that does not need animals to die. Yet, these products and potential products produce a visceral reaction in some people, not (just) because of what they are, but because of what they represent. In a study about acceptance of cultured meat, participants used the words 'artificial, fake, unnatural' most frequently in word associations, but other words included weird, strange, sterilized, disgusting, gross, abomination, dystopia and, for some, ethical.[11] Vegetarians and vegans were hypothetically much *less* likely to try these products according to their own reported feelings about tech-meat, echoing another study which found vegans to be positive about cultured meat, but unlikely to try it.[12] It was the technological aspects and unnaturalness of the products that produced this distaste. So, what is it that produces this ick factor, can it be overcome, or are the cultural dimensions of food, discussed in Chapter 5, just too powerful when it comes to what we want to eat?

Animal products are some of the most unnatural on the planet. Most animal foods that humans eat today would have been virtually unrecognizable throughout most of history. A chicken nugget, a cheese slice or a beef burger are ultra-processed foods, of course, but even the animals they come from have been radically altered through selective breeding, enclosure and, more recently, gene editing. But because they are part of a collective cultural landscape of food and have been on the market for a long time, they aren't exposed to the same scale of disgust as novel cellular meats. Their healthfulness is questioned, but their place on the market is safe. What this emphasizes is that the kinds of food we understand as acceptable is culturally shifting, across both time and space. Things previously or elsewhere common to diets – eyes, hearts or feet, for example – are unusual to others. What we eat isn't just about nutrition, it's about enjoyment, cultural connection, sociability and pleasure. It is also about the kinds of values that a society holds.

Food is contentious and it is political, and access to food has never been equal. It is used as a weapon, its withholding or blockage an exertion of power and strategy of oppression and even warfare. The implementation of hunger as a political tool has countless examples throughout history and continues to the present through physical blockades as readily as it is through economic ones. With the rise of technological agribusiness, the risks of access to food that is healthy, sustainable and culturally acceptable is likely to become more, not less, unequal if food systems power is still held by corporations prioritizing profit. While novel technologies in cellular meat might make dietary transitions easier in allowing people to cook with familiar products, without (most of) the harm to animals, there are also risks of a continued business-as-usual approach to food that doesn't challenge consumption, production or relational aspects of eating. The translation of cultural food ideals from patriarchal, meat-eating society into a vegan future might be enabled by a technical shift that sees cultured meat and precision fermentation produce convincing replicas of animal foods. The technical solution to problems of eating, then, might miss a large part of the picture in the potential of veganism to shift ideals, relationships and connections with the non-human world.

Relationships with nature

According to the Rights and Resources Initiative, a non-governmental organization working to encourage forest tenure and policy reforms based in the United States, as much as 65 per cent of the world's land area is held by communities through customary and community tenure systems,[13] but only a fraction of these are recognized through secure, formal and legal recognition or designation of Indigenous Peoples and local communities. According to the World Bank, 'land is at the center of several development challenges',[14] such as food security, addressing climate change and increasing resilience. But land ownership is concentrating in ever-fewer hands, with the richest 10 per cent of the rural population controlling 60 per cent of land, meaning 2.5 billion people have unequal access to the land they work in small-scale agriculture.[15] Just 1 per cent of farming corporations own 70 per cent of the world's farmland, and this unequal distribution is a driver for hunger and poverty.[16] While the idea of a return to the land might be a romantic and desirable one in future food systems, the question of whether a relationship *to* the land can be transformed is as important as what the land is used for.

In England, 63.1 per cent of land is used for agriculture,[17] while globally, agriculture accounts for around 38 per cent of land use.[18] Advocates of regenerative agriculture and pastoralism argue we should not be moving away from animal farming, but instead looking at shrinking animal agriculture and farming in ways that enhance the land. However, as George Monbiot has argued in *Regenesis*,[19] an extensive agricultural system consisting of pasture farming would require more land than the Earth could ever provide, even if, as pastoralists claim, livestock farming takes place on land that is otherwise unusable. In Kenya, for example, cow farming takes place on the arid and semi-arid lands of the nation, which make up 84 per cent of its total land surface, supporting eight million Kenyans, half of the national livestock and 65 per cent of wildlife.[20] A large portion of this farming is pastoral, meaning cows graze on natural vegetation across a large area, also known as extensive grazing. Despite advocates arguing that this is the best way to produce animal foods, reports from Kenya point towards multiple problems arising from and affecting the system, including soil erosion, natural resource degradation, the inability for pastoralists to move, poverty and prolonged droughts.

Meat consumption is predicted to rise around 30 per cent across Africa by 2030 and with considerable amounts of land already dedicated to agriculture and starting to fail, small-scale pastoralists will be pushed further into poverty. Furthermore, with global environmental concerns looking more closely at the food sector and methane emissions, there is the potential for a two-tier system of food production – and climate-related sanctions – to emerge. While in the United States and Europe, technologies such as food supplements and face masks for cows (that capture methane emitted when cows breathe and convert it into carbon dioxide and water) are being employed to reduce emissions,[21] in places like Kenya that still rely largely on extensive farming systems, their emissions from animal farming are likely to remain high. If carbon budgets and sanctions are introduced, they are far more likely to affect countries like Kenya than those like the United States. In environmental debates, it is largely agreed that eating *less* meat is a universal necessity. However, there remain tense and often heated disagreements over how much less, how that meat should be produced and what global inequalities that shift might create. Technologically produced food, whether of animals or cellular products, is likely to deepen current inequalities in the face of a worsening climate crisis. But a romanticized return to the land is neither practical nor possible, and it also doesn't necessarily shift perspectives of human–nature relationships away from exploitation.

Veganism produces particular visions of the food system (and animal production more widely) that are not only divergent from mainstream visions, but also vary within veganism itself. Broadly, when it comes to food systems, these might be split into two camps: the techno-vegan and the rooted vegan. Techno-veganism believes that the appetite for meat is here to stay and that current food systems (intensive or extensive alternatives) are ecologically disastrous and will kill the planet, if they haven't already done so. They also believe, to at least some extent, in a technological future that can enhance biodiversity and ecologies by removing a proportion of food production from the natural world. This would free up land for animal and human flourishing. Particular critiques of techno-veganism are that it still relies on the use of animals, as cellular-based meat requires donor animals for cells, at least at first. Jan Dutkiewicz and Elan Abrell[22] have made an ethical case for sanctuary models for donor animals, acknowledging that while it isn't ethically satisfactory, these difficult questions must think about principles of least harm, especially in comparison to the industrial agriculture of today, and with the expectation that donor animals would not be needed long term. There are also concerns over capitalist interests and investments that shroud these technologies in uncertainty, on top of concerns over feasibility.

Alternatively, a rooted vision of future vegan food systems, that shares some values with pastoralist movements, albeit without animal agriculture, imagines the human population should be fed not through a severance from the land, but rather a *return* to it. In this vision, feeding humans and animals is also the nourishment of the planet, with kinder, ecologically informed modes of production that aim to help human, plant and animal life thrive. These visions might cite Indigenous, peasant or traditionalist food production as inspirations, with a particular focus on knowledge and land rights. While these small-scale frameworks might be preferable to reconnect us to the land, it's also not clear how feasible this is at scale, particularly given how much more land this would need to produce the world's food and where lots of the planet's soil is already degraded. Rather, this kind of veganism is a set of *principles* that recognizes how a vegan food system is situated within wider complex challenges, including those of hugely increased labour requirements and lower profits. Usually, these agroecological methods continue to centre animal farming, but a vegan vision is also emerging that focuses on plant proteins, such as the vegan agriculture discussed in Chapter 3.

These visions of techno-futures or a return to the land are usually conceived of as opposite ends of the spectrum, irreconcilable. Yet, avid proponents of both share an ethics and passion for a better world for animals (and humans and the Earth). Thinking through veganism as an umbrella principle with different approaches to future food production can offer a novel and realistic vision for the future. For example, cultured meat might target the fast-food industry where animal meat is already hyper-processed and the taste might be easier to replicate and replace, while separate, small-scale and ecologically regenerative agricultural programmes can repair soil health, educate people about where our food comes from and contribute to self-sufficiency. Veganism's vision, then, might not be singular but this isn't necessarily a bad thing. Instead, bringing veganism seriously into future planning allows for debate, difference and a patchwork of opinions and practices that approach food and the environment from a diversity of expertise.

* * *

The current global food system is not working. It is both contributing to climate change and suffering from its consequences, as well as being susceptible to political and economic shocks. Tech-food developments are promising an alternative that could be more resilient and less damaging to the environment, but these technologies aren't currently in a place to be scaled up effectively and affordably. Tech-foods are often focusing on plant-based or cultured products, that would be vegan if brought to the market, but the development of these products isn't only aimed at people who are already vegan. It is about finding ways to navigate – and make money – in the face of food system transition and potential collapse. Nonetheless, the idea of techno-veganism as offering a solution to some of the world's biggest problems is worlds away from veganism as a niche fringe movement, highlighting veganism's potential social, political and economic power.

Novel food products won't cater strictly to vegans; their potential impact is going to be realized in offering a replacement for carbon-intensive and ecologically harmful foods in animal-based diets. The idea of *fake* meat or *alternative* proteins can produce reactions of disgust or even fear, particularly in those who seek a diet that is local, or organic, or supports small businesses, none of which is at odds with other kinds of veganism. There are valid concerns that techno-veganism will be funded by billionaires and companies who already dominate the food market. This kind of technology isn't accessible to just anyone. This perhaps undermines radical vegan politics of justice and fairness, and it's therefore necessary to look beyond techno-veganism to think about and build allegiances with other food practices that might become an umbrella of vegan production serving a range of different desires and needs.

These technological advancements seem to be offering the potential to produce food in way that eliminates the need for animals to be killed for people to eat meat, but technology can't save us from environmental crises and violent relationships with the non-human world unless it also ushers in an ethical shift. Technology might bring about the ethical shift needed for society to update its values; but it also risks a further severing of humans from nature. A view of the food systems that wants to see humanity more connected to the world, not less, is rightly sceptical of these technological promises. But where there is a need for technology, as is arguably the case in the food sector, it's possible that ethical shifts will follow advancements, not precede them. As it stands, techno-veganism is offering a future of consumption that continues contemporary patterns, albeit ones without the environmental footprint and the violent use of animals. Transforming the global food system, built over the last century, may feel impossible, but all hope is not lost. However, it is a future that remains to be seen and cannot ignore the ways the food and technology are bound up with other struggles for justice.

7

JUSTICE

The rise of techno-veganism has brought new practical and ethical questions for vegans, but debates around the boundaries and practice of veganism are nothing new. Sociologists have explored how ethical vegans 'differentiate between those who "eat" vegan and those who "live" vegan' – and how this creates hierarchies among vegans around a kind of purity politics.[1] This can create conflict between vegans and can also result in self-criticism if people feel they can't live up to the standards required to be a 'good' vegan. Questions of authenticity and commitment in veganism are important in understanding how vegans engage with social relationships and their place in the world around them.[2] Vegan ethics and identities are, therefore, not produced on their own: they interact and overlap with other kinds of commitments and priorities. In this chapter, I look at the potential of a veganism focused not just on animals, but on justice for all. To do so, I draw on examples of interconnected vegan practice across different political commitments.

Veganism in practice

While veganism varies across individuals and places, it has often drawn together connected threads of animals, the environment and health to think about the *relationships* between how we consume, and the lives and deaths of others. A core part of this has been the relationship with the self, and in older forms of veganism, a relationship with spirituality. Yet, veganism's more recent iterations have sought to decouple activism from emotions and sentimentality, most famously under the guise of the animal rights movement in the 1970s, and now commonly found in the usage of Richard Ryder's term *speciesism* that seeks to draw parallels between the oppression of animals and the oppression of human groups. While this idea has become common parlance today, it can ignore not only the problematics of drawing comparisons between human and animal suffering, but also overlook work rooted in empathy, emotions and sentimentality, predominantly pioneered by intersectional activists.

The three main priorities of health, animals and the environment have attracted different people to veganism. Some have become vegan to improve their bodies and found themselves increasingly interested in the ethics of eating animals. Others have omitted animals from their diet to save the planet and seen improvements in their health and energy as well as an eased conscience. This has created new tensions in the vegan community, as investment, marketization and capitalist greenwashing have co-opted the label 'vegan'. While veganism has long-standing debates and disagreements, as a fringe movement, these internal nuances were not previously on full display, as veganism was not really taken seriously as a social movement or cultural practice. The growing plant-based market, particularly influenced by its connections with environmental eating, has left critiques of the wider relationships between humans and animals largely untouched. The contemporary growth of veganism, in its focus on diet, has become decoupled from wider ethical changes and is now less concerned with building and practising more equitable ways of living with animals. Veganism in practice, then, does not *necessarily* create new relationships with the non-human world: the work of vegan world-building is a separate endeavour, and perhaps one less open to co-option by capitalism, but therefore one that is often omitted from mainstream discourses.

During my research with vegans, it was interesting to learn that since becoming vegan, the presence of animals in most people's lives had disappeared. Josh, a vegan for ten years, told me:

> I don't have any animals in my life any more and I don't really have any interactions with animals on a day-to-day basis. Because I stopped going to zoos and everything like that a decade ago, I rarely actually see any animals any

more, the only ones that I see on a regular basis are the ones at protests.

But when they did meet animals, another person, Jane, told me that they:

[F]eel like I have got a different relationship with animals than I did before ... when I see cows in the field now, I don't know if I've got a deeper bond with them, but I definitely think about animals more now, they weren't really in my circle of concern before.

While it would be unreasonable to make any assumptions from these quotes that *all* vegans spend less time with animals or feel differently connected to them, what is interesting to draw out is their identification of the lack of ethical spaces to interact with animals in contemporary society. Becoming vegan isn't just about what we eat: it's about a changed way of being in the world. Like many kinds of radical and activist movements, veganism requires a commitment to ongoing education and learning not just about what is or isn't vegan (see Box 7.1), but about the way being vegan shifts relationships with the self, others and the world.

Activism and education are, in some ways, inseparable. For vegans, their education does not end once they make the decision to go vegan, but involves a life-long commitment to education, both self-education and education of others. A Community Interest Company called Veganism in Education

Box 7.1: Veganism versus ...

From fish in a favourite wine, to the use of milk in condoms, bugs in make-up, and bones in sugar, veganism reveals how our everyday lives are intimately reliant on animals in what we eat, wear and use. Even the plants that we eat rely on forms of harm, whether field mice being killed in the harvest, the clearing of land for agriculture affecting the habitats of birds or pesticide spraying killing insects and other animals.

While vegan consumption has become easier with growing awareness and availability of food and lifestyle products, such as cosmetics, cleaners and clothes, it still requires a conscious effort to ensure things are vegan, particularly when looking out for human exploitation in production and the sustainability of products. Veganism, as I wrote in the first half of this book, is not one practice or belief. It has different motivations and commitments. It is therefore not always the case that vegan ethics and practice align with other social and political values.

If, for example, new research was published tomorrow that found that eating cows was, after all, environmentally friendly (with the caveat that no such piece of research could exist, no single study could make such a claim and the scientific community would have serious questions about its validity), would I rush out and scoff down a cow? No! Of course not, and neither would many other vegans, as our commitment to animals outweighs that to dubious environmental claims. We'd find another way to reduce our footprint on the planet.

There are inherent cruelties in any consumption relationships in the contemporary world: almost nothing is equal or fair. To the despair of many, including myself, the suffering, pain and exploitation of both humans and animals is today as bad as it has ever been. There is no 'cruelty-free' way to exist in this system. So, no, veganism isn't cruelty free, but this isn't a 'gotcha' that proves vegans are wasting their time. It's rather part of a conversation around how multiple, intersecting issues not just can, but *must*, be addressed together.

Perhaps the writer and animal advocate Brigid Brophy put it best, as she often did, when she wrote in 'The Rights of Animals'[3] in 1965: 'where animals are concerned humanity seems to have switched off its morals and aesthetics—indeed, its very imagination'. Only by reviving this imagination to understand, and challenge, contemporary economic and political systems can a veganism beyond cruelty be established.

(VinE) was founded in 2021 to support professional educators to encourage humane education across the life course. Where children's education, particularly in their younger years, often relies on romanticized visions of our relationships with animals, from books that show idyllic farm life to the provision of dairy milk in schools and nurseries, VinE offers alternative resources and provides teacher training. Vegan education isn't always formal, it also happens outside mainstream institutions and, for many of us, much later in life. For example, vegetarian cafes and local chapters of activist networks have long hosted local meet-ups, organizational groups and outings for vegans, far pre-dating the now-common online spaces of support. These often overlap, particularly in the past, with other subcultures such as environmentalists or punks. Today, being vegan, while still stigmatized, is definitely more common and visible than it was even ten years ago. However, vegan-centred community remains important to education, connections and debates over the future of veganism – and what justice might mean for humans, animals and the planet.

Veganism and ...

Veganism has connections with many other social movements. Ecofeminists like Carol J. Adams and Black vegan-feminists like Aph and Syl Ko and Breeze Harper have long been leading the charge for recognizing, understanding and resisting shared logics between animal and human oppression. While making comparisons between human and animal suffering has been critiqued for reproducing harmful stereotypes, these thinkers, writers and activists instead point out how systems and structures of suffering operate similarly between different kinds of discrimination and species. For example, the geographer Karen Morin in her book *Carceral Space, Prisoners and Animals*[4] shows how industrial farms, zoos, slaughterhouses and prisons for humans all share particular characteristics of captivity and confinement. She shows that the physical buildings of prisons and industrial agricultural farms are near-identical from aerial views; that solitary confinement and the zoo cage echo one another; that death in prison execution chambers and the slaughterhouse follow similar regulations of separating life from death; and both prisoners and animals are victims of pharmaceutical lab testing. These spaces are not the same, but they follow similar procedures and *logics*.

Contemporary veganism is increasingly nurturing these connections with feminist, anti-racist, disabled and LGBTQIA+ rights movements as part of an interconnected approach to justice. In this penultimate chapter of the book, I offer case studies of activists, theorists and advocates who are working across struggles, leading to reflections on what a holistic vision of justice might look like across species.

Vegan-feminism

In 1990, Carol J. Adams published her book *The Sexual Politics of Meat (TSPoM)*,[5] in which she argues that the oppression of animals and domination of women result from shared processes of both women and animals being made into 'absent referents' in culture and society. Over the last 35 years, Adams and other vegan-feminists have showed how the consumption of animals and the consumption of women use processes of objectification, divided into parts, and subject to violence – both physical and psychological. The association, for example, of meat and masculinity is the foil to the connected violence towards women and animals. Importantly, the actual forms of violence are not identical but echo one another and reveal how patriarchy and meat-eating are intimately connected.

Vegan-feminist work has looked at how meat-eating and masculinity are culturally connected; at animal abuse as part of domestic violence; and milk and egg industries' exploitation of reproductive systems and reliance on sexual abuse. Vegan feminism also emphasizes that most animal activists are women, but these same women are rarely seen as 'leaders' in the animal movement, as pointed out by Corey Lee Wrenn in her work on the Vegan Feminist Network. This online network seeks to rectify this invisibilization of women's labour but also to act as a safe space and place for outreach that recognizes how gender and species are intersecting oppressions in society, politics and culture. Vegan feminism therefore brings *species* into the lens of understanding structural oppression (see Figure 7.1).

Writing in 2010,[6] Carol J. Adams said of *TSPoM*: 'What an optimist I was! I hugely underestimated the antifeminist world – the world vested in women's inequality – and the meat-eating world, and the way they would intensify their interactions.' With a rise in right-wing politics, misogyny and climate change denial, the connection between veganism and feminism is one that is still very much needed.

Black veganism

The idea that our identities are multiple and compounding, and that our identities shape our

Figure 7.1: Feminists for Animal Liberation

Feminists for Animal Liberation: An Ecofeminist Alliance, pictured here, was founded in California in 1981 and was active into the 21st century. They focused on consciousness-raising in both feminist spaces and the animal advocacy movement, as well as with the general public. In their own words, 'they knew that it was not enough to claim an abstract respect for animals; they knew they must embody that respect in their daily lives' (http://www.farinc.org/FAR_home_contd.html).

worlds, is a powerful one. The use of intersectionality as an explanatory and analytical framework has gained popularity over the past few years, with the once-academic notion of intersectionality being picked up by activists and influencers alike. Importantly, intersectionality is rooted in Black feminist legal thought, first coined by US scholar and civil rights advocate Kimberlé Crenshaw, who argued in 1989 that the law's narrow view of discrimination created single-issue analyses that did not fully represent the world. Using examples from case law, Crenshaw argues that the law failed to consider how gender-based discrimination and race-based discrimination *intersect*, and therefore that the law ignored that Black women were both Black *and* women and faced discrimination on both counts. It is important to remember and acknowledge the roots of intersectionality in Black feminist legal theory, to centre race when extending these analyses, as some vegans have argued should be done with regards to species.

In their book *Aphro-ism*,[7] Aph Ko and Syl Ko write about Black veganism and pop culture but argue that intersectionality doesn't go far enough as it 'doesn't really trouble the systems looming over us that we never created. Intersectionality maps out the world that has been *imposed* on us; it doesn't begin the process of mapping out the future' (emphasis in original). This is echoed by activist and vegan farmer Nassim Nobari,[8] who wrote in 2023 that vegan intersectionality has become part of neoliberal capitalism in the vegan movement, and that 'intersectional veganism was never given a clear and consistent definition by its proponents, nor was the term "intersectionality" used in a manner consistent with how Kimberlé Crenshaw defined it'. Black veganism counters some of the problematics of mainstream veganism, but also situates veganism as collaborative with wider issues of labour, food justice and anti-racism.

Class and veganism

In its recent history, veganism has been caricatured as a White, middle- and upper-class pastime, reserved for wealthy celebrities and those who can afford to shop in specialist stores. The growth of corporate veganism has come with an increased accessibility, as not only are supermarkets stocking more vegan foods, but big brands are cashing in on the plant-based market, primarily in the food sector, but also in clothes and cosmetics. But the connections between class and veganism aren't only about capital and purchasing power, they also raise important questions in thinking about animals and labour rights.

American anthropologist, Angela Stuesse, has undertaken a six-year ethnography on the lives of migrant workers in Mississippi who work on intensive chicken farms. Her book, *Scratching Out a Living*,[9] looks at the dangerous work, the hard and badly paid jobs and the health impacts of working on such farms, which include not just psychological distress, but respiratory problems from inhaling ammonia. The exploitation of humans and animals in animal agriculture has led to arguments from socialist vegans that veganism should be a leftist practice as it enables new forms of solidarity and challenging harmful social and economic systems.[10]

While class and veganism have often been talked about in a dismissive way, there are vegan movements and projects that seek to build connections with these people suffering within the animal industry. Writing in 2023, Andrew Gough[11] argued that extending empathy and understanding to slaughterhouse workers could be important to vegan activism – not least because these workers are often precarious migrant labourers with little choice in the work they do. Slaughterhouse workers are often badly paid, insecurely employed and suffer from post-traumatic stress disorder as well as physical health problems because of their work. Accounts of slaughterhouse work are often harrowing and the violence workers undertake in the slaughterhouse often has ramifications for substance abuse, domestic violence and crime rates. Vegan scholar-activist Kim Stallwood credits his summer training course at a chicken factory with pushing him to become vegetarian due to the horrors he witnessed, but for precarious human labourers, there aren't always more choices open.

Queer veganism

In *A Queer Vegan Manifesto*,[12] Rasmus Simonsen argues that 'the promise of becoming vegan ... is to challenge, or queer, always and everywhere, the normative demands that are placed upon our genders, sexualities, and diets'. Veganism as a queer theory and practice challenges normative society that centres anthro-privilege (the privilege of humans) and by refusing cultures of meat-eating, veganism can trouble ideas of, for example, the family, reproduction and masculinity, while also offering a 'titillating approach to life'.[13]

Queer veganism is not simply the practice of veganism by LGBTQIA+ people; it is itself a queer practice that challenges dominant norms and assumptions about what we consume – and how. Carrie Hamilton,[14] vegan writer and scholar, has written about how memory, queer identity and veganism intersect in the mourning of the use of leather by queer vegans in leather subcultures (a queer subculture who organize around sexual activities including leather). Hamilton writes that while she feels a sense of belonging in the subculture, she also feels alienation and ambivalence that has become common in her vegan life.[15] Materiality and objects are part of community-building and, as Hamilton's work shows, this doesn't necessarily flow easily into every part of queer life.

Queer theorists Emelia Quinn and Benjamin Westwood[16] have also written on the alignment of veganism and queerness as challenging normative identities; expanding queer ideas to animals; and offering new possibilities for queer veganism to together rethink futures. In an essay on his own veganism, vegan scholar Robert McKay recounts the homophobia and vegan-phobia he has experienced by eating disruptively, and vegan men have reported experiencing homophobia because of their diet.[17] The association between veganism and femininity has led to prejudice and homophobia towards vegan men, but it also paves the way for more fluid and subversive forms of masculinity and sexuality.

Disability and veganism

Some utilitarian figures in animal rights, notably Peter Singer in *Animal Liberation*,[18] have been rightly admonished for comparing the rights of animals to the rights of disabled humans. Yet, establishing empathies between animal and disability liberation can, as Sunaura Taylor shows in her book *Beasts of Burden*,[19] challenge the separation of human from animal in a process of 'cripping animal ethics' that doesn't rally against conversations around disability and animality but instead connects oppressions beyond species. In her own words, 'this connection did not lie, as many people suggested, in my being confined to my disabled body, like an animal in a cage. Far from this, the connection I found centred on an oppressive value system that declares some bodies normal, some bodies broken, and some bodies food.' Taylor argues that the root of resistance to comparisons with animals is discrimination against animals – despite how varying sentient beings might be 'aware, sentient, or intelligent'.

While veganism and disability justice can have empathetic and productive overlaps, this doesn't necessarily translate to vegan spaces of activism. Animal rights and disability activist Michele Kaplan has spoken about ableist discourses in online vegan spaces, where being vegan is sometimes presented as a fix-all 'cure' for any illness. Kaplan spends time apologizing for online vegan discourses that leave disabled people feeling unheard or dismissed; yet she doesn't believe that veganism is inherently ableist – even if there are sometimes ableist discourses: 'Just as a disabled person, I don't exist to be someone's inspiration nor target of pity, animals do not exist to be our meals and clothing.'[20]

Thinking about disability and veganism isn't limited to human connections; many animals are disabled by the farming, entertainment and fashion industries. Farmed animals who have been intensively selectively bred often suffer from deficiencies and genetic traits that mean they can't survive. Vegan sanctuaries sometimes step in where they can, to rescue and care for these animals who otherwise would be headed straight for the slaughterhouse. The LGBTQIA+ run vegan VINE sanctuary in Vermont, for example, is home to a multispecies community that doesn't separate animals by species and is home to hundreds of animals rescued from factory farming.[21] Focusing on interlocking and intersectional analyses expanding from veganism in farmed animal sanctuaries, VINE provides a space which enables animals to learn to be themselves without limit.

Veganism has been related to and practised in conjunction with radical social justice movements. Intersecting logics of oppression have enabled members of marginalized human communities to recognize and act in solidarity with non-human animals. Thinking about the future of veganism, these committed activist pasts will be key to advocating for animals and veganism – although they seem worlds away from contemporary debates focused on, for example, technological eating.

Veganism and struggle

Liberationist and hopeful visions of veganism as a space of justice are not the only kinds of experiences that vegans go through in their transition. Vegan geographer Richard J. White[22] has written about how his engagement with the world around him changed when he went vegan and began to see signs of interspecies violence everywhere, 'hidden in plain sight', in the shops, restaurants and city he walks through on his daily commute. While the city might seem to present few obvious sites of violence against animals, it is actually built on the pain of animals. The contemporary city is hostile to most forms of animal life, having pushed out animals from within its boundaries in processes of sanitization, while feasting on ever more of their bodies as food and clothes. Urban life isn't just hostile; it is actively built on violence for consumption, growth and human life. Rural areas don't fare much better, with farms and slaughterhouses often situated away from dense areas of human habitation. In these landscapes, not only are the violences of industrial farming hidden from most people's sight, but the slow violence of environmental damage surrounds these landscapes. Becoming vegan opens the eyes of people to how spaces can be sites of harm that can reshape engagement with the world around us in negative ways.

It isn't just physical places that can invoke this sense of emotion as people become vegan; social and personal relationships also experience shifts, and these can be painful (see Chapter 1). Stories of alienation, ending of relationships and loneliness are common for vegans whose families and friends might not accept this new version of them. Even as people feel that they are moving to align their practices with their values, the path after transitioning to veganism is not a smooth one, with social stigma being cited as a reason that people stop being vegan.[23] While veganism can offer social connection through online support groups, and for the development of new identities that enable people to practise their values, it can also lead to stigma and isolation from their previous life. Presenting a version of veganism that is entirely liberatory and celebratory fails to account for these negative experiences.

Veganism is never just one thing. It can intersect with other social justice movements, with radical ways of life and with new senses of identity and embodiment. These pathways through veganism and its intersecting identities and practices, though, are not without struggle and difficulty. Even within communities of vegans, some people are more likely to experience stigma, oppression and alienation; mainstream veganism can be uncomfortable for Black vegans; and the greenwashing of veganism is producing alienation for people who have been vegan for a long time and associate their veganism with political ethics. Similarly, while around three-quarters of vegans are women, many of the self-appointed leaders and mouthpieces of veganism are White men, and their use of harmful comparisons between women and animals, race and animals, and disability and animals are still shockingly commonplace in their vegan activism. Vegan spaces do not, therefore, de facto act as sanctuaries from negativity in social and physical non-vegan spaces; they can also be riddled with the same forms of hierarchy and domination.

Veganism seeks out a way of living that reduces and seeks to eliminate as far as possible harm to other living beings. For many vegans, this includes human beings, as evidenced in the rich and vital work of the vegan work showcased in this chapter. For these activists and scholars, veganism isn't a single issue that distracts from other social and political justice movements; it is deeply ingrained in and essential to resisting oppressive logics and structures that act in similar ways across different of groups of people, animals and environments. As veganism has become a more mainstream movement, it has become harder to resist and push against practices of veganism that solely focus on dietary change or consumption. Learning from and centring activists and thinkers who conceptualize and practise veganism as liberatory politics, the future of veganism begins to seem much harder to capture and greenwash.

* * *

To end this chapter, I want to share a quote from ecofeminist Lori Gruen from a book called *Ecofeminism*, which she co-edited with Carol J. Adams:

> Living today, even for vegans, involves participating unwittingly in the death of sentient individuals. We can rail against the massive violence that is done to the huge number of living beings who did nothing to deserve their tragic fate, but our political commitments and moral outrage doesn't clean our hands. We harm other (human and nonhumans) in all aspects of food production. ...
> Vegan diets are less harmful than those that include

animal products, to be sure, but the harms and deaths occur nonetheless.[24]

The process of becoming vegan invites people to learn more about the world around them, and the things that they eat and use in their everyday life and where they come from. This can cultivate particular ethics and practices, and it is this wider shift in consciousness that is often considered 'true' veganism. But, increasingly, veganism is being conflated with eating and consumption practices, and people following a plant-based diet are as resistant to being cast under a wider vegan ethics as some vegans are to veganism being cast as *just* a diet. Popular understandings of veganism do, largely, focus on it as a diet, but the complexities of defining veganism are as hard as they have ever been, and perhaps even more so today as veganism is growing. It's clear that there is no one singular vision or practice of veganism. It encompasses a range of ideas, has many complex and long-standing debates, and is constantly changing as more people join the movement and bring with them their own perspectives and ideas.

It is therefore vitally important that veganism can be enabled to flourish and develop along multiple values in order to fully reach its potential, particularly in connection with other social justice movements. While vegans may be notorious for internal debate and disagreement, it is the richness and passion of these very same debates that have allowed veganism to incorporate and envelop a diversity of perspectives.

Veganism will always be more than a diet, because of the values and practices of vegans in creating and connecting struggles together. Looking, then, to what veganism stands for is complicated, but it also allows for the imagination of a future very different to the present that is both realistic and hopeful. Veganism that embraces and upholds other social and political movements is, therefore, *for* a just future for all.

8
CONCLUSION: VEGAN FUTURES

Veganism's connection with other social movements shows its potential for a vision of justice that wants to transform the world. But this vision is a long way from the veganism that has swept society, which has undoubtedly reproduced ideas that individual consumption and action define vegan practice. This individualization of veganism is linked to its uptake and greenwashing by producers and corporations who are looking to expand their markets. But there are also vegans who buy into this ideology, and do not see veganism as part of a social justice struggle or political movement. So, what, if anything, brings veganism together into a unified vision, or is it impossible to overcome these divisions?

Veganism, in all its forms, represents a changed relationship with the self, with society, with animals and/or with the environment. This transformation is by definition engaged with hope and desire to create a better future. While this may stop at diet or consumption for some, overall, veganism is expansive and ranges from the relatively conservative to the politically radical. It is these very differences and divergences that have produced and allowed veganism to grow, change and be challenged by new people coming to the movement, and in turn to have more research and evidence about the wide-ranging benefits of becoming vegan. Veganism may begin with an individual choice, but it also requires a fundamental shift in how people think about food, and their relationships with animals and the natural world. By encompassing different perspectives and practices, the vegan community, broadly defined, is often seen as over-complicated or at odds with itself, which can be cast as a negative – particularly in relation to internal disagreements. I believe that this is actually a strength of veganism, that it can take myriad forms, be culturally specific, evolving over time to encompass new ethics and practices.

Looking to the future is key for veganism; becoming vegan can be seen as investment in that future, in making it better – whether for personal health, the environment, animals, or all three. This doesn't mean that understanding its histories and successes is unnecessary. In fact, a better understanding of how veganism has evolved and grown as part of society, but also in contention with it, is critical to maintaining the momentum that it has recently gained. Animal activist groups are being founded at an incredible rate across the globe, taking up locally specific causes to advocate for animals, local people and environmental causes. Plant-based diets are now being adopted as part of a political and ethical movement, and vegan chefs and cooks are experimenting with plants to recreate traditional cultural foods in vegan forms, making it more accessible. In making versions of fish and chips, sushi, or piri-piri chicken out of plants, not only are vegans innovating and celebrating the versatility and excitement of learning to cook plants, but are also able to hold on to their local traditions and comfort foods. This global respect for veganism is something that, frankly, feels both impossible *and* inevitable for long-term vegans.

I want to end by returning to this book's key question: *what is veganism for?* Veganism is a way of living that aims to improve animals' lives, human health and the planetary future. It has emerged across the world in different forms and goes beyond simply a change in diet; becoming vegan changes the way people engage with the world. For many vegans, there are strong connections between their vegan ethics and practices and other social issues. In the recent surge of veganism, intersectionality is becoming an increasingly important feature, with the connections between humans, animals and the planet being seen as increasingly interconnected and requiring holistic collaborative solutions. Veganism is multiple and complex, home to a whole range of perspectives, from those who want us to live a more connected life in nature, to those for whom technological advancements offer an exciting future. Throughout this book, it has been my goal to balance the terrifying crisis and state of the world today with hopeful visions for the future; not just what veganism is against, but what it offers. I am therefore going to finish with a speculative experiment, laying out what a vegan world could look like, and inviting you to leave this book and think about how you might work towards or imagine that future world in your everyday life, practices and relationships with others.

In his 2017 mockumentary, *Carnage*,[1] British comedian and vegan Simon Amstell sets the scene in 2067, when the world has become vegan. Older generations are struggling with the guilt of their past animal eating, while young people can't understand why their parents and grandparents would have been involved in such violence. The mockumentary moves from the 20th-century history of veganism, into the then-present of 2017 that looks much like today – animals being promoted on TV cookery shows, veganism being promoted by some celebrities but mostly ignored, as well as the effects of climate change and illness linked to meat consumption. In the film, technology plays an important role in the shift from meat-eating society to vegan norms, first through the development of a Thought Translator that shows people what animals are feeling, and then through virtual-reality technology in 2067 that allows young people of the future to see what the pre-vegan world was like, leading one to ask in distress, 'Why would anyone eat a baby? It's just a little baby.' More recently, George Monbiot released a video with WePlanet (then named RePlanet Media),[2] in which he asks, instead of being disgusted by new food technologies, 'what if it was in fact the other way around? Imagine we were already doing that ... and someone came along and said, "let's eat animals instead". EURGH animals!' This mobilizing of alternative lives and worlds recasts the future not as an inevitable one of collapse and disaster, but instead still open to the possibility of a radically different, and better, world.

The path to that world may not be clear, yet, but I can see it on the horizon. And so can millions of others.

If humans were to stop eating animals, everything would change, for us and for them. But this isn't a simple thought experiment: there are *already* people and animals living in a post-meat world, spaces of multispecies cohabitation can be found in places from VINE sanctuary in Vermont, a queer-run animal refuge; Jacobs Ridge, an animal sanctuary in Murcia, Spain; Sadhana Forest Animal Sanctuary in India; and even in our own back gardens, which might be shared with birds, squirrels, insects and plants. If veganism has to articulate, then, what will happen to the animals if we stop eating them, especially those that may not be used to a life outside of human domination, the answer is simple: if the world goes vegan, our relationships to animals will have transformed so drastically that their longevity, safety and happiness would no longer be under threat. If we all go vegan, there won't be a loss of animal life on Earth, but an abundance; and this life would be valued, protected and flourishing. What is veganism for, then? Most of all, veganism is for the future.

NOTES

Prologue

[1] Eva Giraud (2021) *Veganism: Politics, Practice, Theory.* Bloomsbury.

[2] Catherine Oliver (2021) Vegan world-making in meat-centric society: the embodied geographies of veganism. *Social & Cultural Geography*, pp 1–20.

Chapter 1

[1] Marie Kacouchia in an interview with Lizzy Davies (2023) Rainbow plates: the chefs reawakening Africa's taste for vegan food, *The Guardian*, 27 February. https://www.theguardian.com/global-development/2023/feb/27/africa-taste-for-vegan-food-plant-based

[2] Nikita Singh (2020) Veganism isn't new for Africans—it's a return to our roots, say these chefs and entrepreneurs. *ProVeg*, 27 May. https://proveg.com/za/2020/05/veganism-isnt-new-for-africans-its-a-return-to-our-roots-say-these-chefs-and-entrepreneurs/

[3] Molly Long (2022) How many vegans are there in the UK? *Food Matters Live*, 18 May. https://foodmatterslive.com/article/how-many-vegans-are-there-in-the-uk/

[4] https://www.globenewswire.com/news-release/2022/03/30/2412865/0/en/Europe-Plant-Based-Food-Market-Worth-16-7-Billion-by-2029-Exclusive-Report-by-Meticulous-Research.html

[5] https://www.new-nutrition.com/keytrend?id=135

[6] https://marketingtochina.com/vegan-movement-china-big-appetite-veganism/#:~:text=Reports%20indicate%20that%205%25%20of,diseases%20and%20cut%20carbon%20emissions

[7] Jeffrey M. Jones (2023) In U.S., 4% identify as vegetarian, 1% as vegan. *Gallup*, 24 August. https://news.gallup.com/poll/510038/identify-vegetarian-vegan.aspx

8   Aashi Keshair (2021) Is vegan food successfully penetrating the Indian market? 1 October, https://bwhotelier.businessworld.in/article/Is-vegan-food-successfully-penetrating-the-Indian-market-/01-10-2021-406974/

9   All statistics included have been sourced from The Vegan Society, who have collated statistics on veganism across the world here: https://www.vegansociety.com/news/media/statistics/worldwide

10 https://mercyforanimals.org/blog/what-motivates-people-vegan/#:~:text=In%202019%2C%20vegan%20travel%20blog,and%2010%25%20cited%20environmental%20concerns

11 Sally Ho (2021) 44% of Hong Kong people who are changing their diets are doing it for animals, survey data shows. *Green Queen*, 19 February. https://www.greenqueen.com.hk/44-of-hong-kong-people-who-are-changing-their-diets-are-doing-it-for-animals-survey-data-shows/

12 https://faunalytics.org/a-summary-of-faunalytics-study-of-current-and-former-vegetarians-and-vegans/

13 Veganuary data, 2014: https://veganuary.com/behind-the-scenes-at-veganuary/; Veganuary date, 2023: https://veganuary.com/veganuary-2023-breaks-all-records/; Veganuary date, 2022: https://veganuary.com/wp-content/uploads/2022/03/THE-OFFICIAL-VEGANUARY-2022-PARTICIPANT-SURVEY.pdf

14 Hannah Ritchie, Pablo Rosado and Max Roser (2023) Meat and dairy production. *Our World in Data*. https://ourworldindata.org/meat-production

15 Caitlin Dewey (2017) Cattle have gotten so big that restaurants and grocery stores need new ways to cut steaks. *The Washington Post*, 7 December. https://www.washingtonpost.com/news/wonk/wp/2017/12/07/cattle-have-gotten-so-big-that-restaurants-and-grocery-stores-need-new-ways-to-cut-steaks/

16 Dan Blaustein-Rejto and Alex Smith (2021) We're on track to set a new record for global meat consumption. *MIT Technology Review*, 26 April. https://www.technologyreview.com/2021/04/26/1023636/sustainable-meat-livestock-production-climate-change/

17 https://www.fao.org/sustainability/news/detail/en/c/1274219/

18 https://www.vegansociety.com/go-vegan/definition-veganism

Chapter 2

1   *China Agricultural Outlook Report (2022–2031)*, China Agricultural Outlook Conference, 12 July 2022; *OECD-FAO Agricultural Outlook 2022–2031*, OECD, 29 June 2022.

2   Oliver Milman and Stuart Leavenworth (2016) China's plan to cut meat consumption by 50% cheered by climate campaigners. *The Guardian*, 20 June. https://www.theguardian.com/world/2016/jun/20/chinas-meat-consumption-climate-change

3   Marcello Rossi and Undark (2018) China's love for meat is threatening its green movement. *The Atlantic*, 31 July. https://wildaid.org/chinas-love-for-meat-is-threatening-its-green-movement/

4   You can see examples in the United Kingdom at Beef Farming Technology Projects, Department of Agriculture, Environment and Rural Affairs, https://www.daera-ni.gov.uk/articles/beef-farming-technology-projects

5   Stephanie G. Schuttler, Kathryn Stevenson, Roland Kays and Robert R. Dunn (2019) Children's attitudes towards animals are similar across suburban, exurban, and rural areas. *PeerJ*, 7: e7328. https://www.ncbi.nlm.nih.gov/pmc/articles/PMC6659664/

6   See https://europenowjournal.org/2018/09/04/looking-backward-moving-forward-articulating-a-yes-but-response-to-lifestyle-veganism/

7   For readers interested in animal rights, Steve Cooke's book in this series, *What Are Animal Rights For?* will be essential reading.

8   Peter Singer (1982) The Oxford vegetarians: a personal account. *International Journal for the Study of Animal Problems*, 3(1): 6–9.

9   Carol J. Adams and Lori Gruen (2014) *Ecofeminism: Feminist Intersections with Other Animals and the Earth*. Bloomsbury, p 1.

10 More information about food production can be found here: https://www.oecd-ilibrary.org/sites/aa3fa6a0-en/index.html?itemId=/content/component/aa3fa6a0-en#section-d1e19629

11 Carol J. Adams (1990) *The Sexual Politics of Meat*. Bloomsbury, p 192.

12 You can view the photograph on the Natural History Museum's website at https://www.nhm.ac.uk/wpy/gallery/2021-hope-in-a-burned-plantation. It was shortlisted for the People's Choice award in 2021.

Chapter 3

1   IPCC Special Report on Climate Change and Land, led by researchers across the world. It can be found at: https://www.ipcc.ch/srccl/

2   As reported in the a UN report published in 2006 called *Livestock's Long Shadow*, which can be accessed here: https://www.fao.org/3/a0701e/a0701e00.pdf

3   https://www.un.org/sg/en/content/secretary-generals-statement-the-ipcc-working-group-1-report-the-physical-science-basis-of-the-sixth-assessment

4   As discussed on BBC Radio 4 in 2023: https://www.bbc.co.uk/programmes/m001sth9

5   https://www.peta.org/about-peta/faq/how-does-eating-meat-harm-the-environment/

6   Fiona Harvey (2021) Global salmon farming harming marine life and costing billions in damage. *The Guardian*, 11 February. https://www.theguardian.com/environment/2021/feb/11/global-salmon-farming-harming-marine-life-and-costing-billions-in-damage

7   Andrew Wasley, Alexandra Heal and Fiona Harvey (2019) Ammonia pollution damaging more than 60% of UK land – report. *The Guardian*, 18 June. https://www.theguardian.com/environment/2019/jun/18/ammonia-pollution-damaging-uk-land-report

8   Center for Biological Diversity (nd) *Chicken v Wildlife: The Environmental Costs of Eating Poultry*. https://www.biologicaldiversity.org/takeextinctionoffyourplate/pdfs/ChickenFactsheet.pdf

9   As reported by *FarmWell*, an online platform serving the UK farming community: https://farmwell.org.uk/seasonal-workers-and-modern-slavery/#:~:text=Seasonal%20working%2C%20modern%20day%20slavery,10%2C000%2D13%2C000%20people%20being%20exploited

10   H. Charles J. Godfray et al (2018) Meat consumption, health, and the environment. *Science*, 361(6399): eaam5324. https://www.science.org/doi/10.1126/science.aam5324

11   And will into the future, 'unless major technological changes disproportionately target animal products' as written by Joseph Poore and Thomas Nemecek (2018) Reducing food's environmental impacts through producers and consumers. *Science*, 360(6392): 987–992. https://www.science.org/doi/10.1126/science.aaq0216

12   Mona Seymour and Alisha Utter (2021) Veganic farming in the United States: farmer perceptions, motivations, and experiences. *Agriculture and Human Values*, 38(4): 1139–1159. https://doi.org/10.1007/s10460-021-10225-x

13   Known as the Stockfree Organic Standards: Ulrich Schmutz and Lucia Foresi (2016) Vegan organic horticulture – standards, challenges, socio-economics and impact on global food security.

III International Symposium on Organic Greenhouse Horticulture, 1164: 475–484.

[14] Jeremy Sharon (2020) Vertical farm produces kosher certified no-insect produce. *The Jerusalem Post*, 29 December. https://www.jpost.com/diaspora/vertical-farm-produces-kosher-certified-no-insect-produce-653635

[15] https://www.thegrocer.co.uk/mergers-and-acquisitions/saputo-snaps-up-scottish-vegan-cheese-brand-sheese/656459.article#:~:text=Canadian%20dairy%20giant%20Saputo%20has,Scottish%20vegan%20cheese%20brand%20Sheese

[16] Denise Chow (2020) Earth Day at 50: why the legacy of the 1970s environmental movement is in jeopardy. *NBC*, 22 April. https://www.nbcnews.com/science/environment/earth-day-50-why-legacy-1970s-environmental-movement-jeopardy-n1189506

[17] Joseph Poore and Thomas Nemecek (2018) Reducing food's environmental impacts through producers and consumers. *Science*, 360(6392): 987–992. https://www.science.org/doi/10.1126/science.aaq0216

[18] Hannah Ritchie (2021) If the world adopted a plant-based diet we would reduce global agricultural land use from 4 to 1 billion hectares. *Our World in Data*, 4 March. https://ourworldindata.org/land-use-diets

[19] Although a totally vegan and organic food system would not be able to provide enough nutrients and calories. The study was undertaken in Applied Computational Life Sciences at Zurich University of Applied Sciences: https://digitalcollection.zhaw.ch/handle/11475/24859

[20] https://veganaustralia.org.au/research/

[21] https://www.theccc.org.uk/publication/land-use-policies-for-a-net-zero-uk/

[22] Maarten Elferink and Florian Schierhorn (2019) The dormant breadbasket of the Asia-Pacific. *The Diplomat*, 12 February. https://thediplomat.com/2019/02/the-dormant-breadbasket-of-the-asia-pacific/

Chapter 4

[1] You can watch Macka B's music on his YouTube channel: https://www.youtube.com/@OfficialMackab

[2] As reported in an NSF commissioned report, 'Food trends, changes and challenges', which can be read at: https://nsfinternational.widen.net/s/fss72qqxsv/ct_food_trends_insight_ereport

3   Hannah Landecker (2013) Postindustrial metabolism: fat knowledge. *Public Culture*, 25(3): 495–522, at 495. https://doi.org/10.1215/08992363-2144625

4   Frances Moore Lappé (2011) *Diet for a Small Planet: The Book That Started a Revolution in the Way Americans Eat*. Ballantine Books.

5   Jennifer L. Bedford and Susan I. Barr (2005) Diets and selected lifestyle practices of self-defined adult vegetarians from a population-based sample suggest they are more 'health conscious'. *International Journal of Behavioral Nutrition and Physical Activity*, 2(1): 1–11. https://ijbnpa.biomedcentral.com/articles/10.1186/1479-5868-2-4

6   Amy Buxton (2022) A new study linked meatless diets with depression, but experts have questions. *Plant Based News*, 18 October. https://plantbasednews.org/lifestyle/depression-meat-free-diet-connection/

7   Matthew Cole and Karen Morgan (2011) Vegaphobia: Derogatory discourses of veganism and the reproduction of speciesism in UK national newspapers. *The British Journal of Sociology*, 62(1): 134–153.

8   Rebecca Gregson, Jared Piazza and Ryan L. Boyd (2022) 'Against the cult of veganism': unpacking the social psychology and ideology of anti-vegans. *Appetite*, 178: 106143. https://doi.org/10.1016/j.appet.2022.106143

9   South African academics cited in Jane Dalton (2021) Meat-eating creates risk of future pandemic that 'would make Covid seem a dress rehearsal', scientists warn. *Independent*, 30 January. https://www.independent.co.uk/climate-change/news/meat-coronavirus-pandemic-science-animals-b1794996.html

10 UN (2020) COVID-19 dress rehearsal for world of challenges to come, Secretary-General tells General Assembly. UN press release following a UN Secretary-General António Guterres' address to a General Assembly in New York. *UN Meetings Coverage and Press Releases*. https://press.un.org/en/2020/sgsm20267.doc.htm

11 Thijs Kuiken and Ruth Cromie (2022) Protect wildlife from livestock diseases. *Science*, 378(6615): 5. https://doi.org/10.1126/science.adf0956

12 David Quammen (2023) Why dead birds are falling from the sky. *New York Times*, 23 April. https://www.nytimes.com/2023/04/23/opinion/bird-flu.html?smid=tw-share

13   Frédéric Keck (2020) *Avian Reservoirs: Virus Hunters and Birdwatchers in Chinese Sentinel Posts.* Duke University Press.

14   Claire Hamlett (2021) The other pandemic: Avian flu is spreading around the globe. *Sentient Media*, 19 January. https://sentientmedia.org/the-other-pandemic-avian-flu-is-spreading-around-the-globe/

15   Claire Hamlett (2022) What this Guardian article got wrong about avian flu. *Sentient Media*, 9 June. https://sentientmedia.org/the-correction-what-this-guardian-article-got-wrong-about-avian-flu/

16   Niñoval F. Pacaol (2023) Veganism and public health: a brief manifesto. *Journal of Public Health*, 45(2): e367. https://doi.org/10.1093/pubmed/fdac138

17   Richard Horton, Robert Beaglehole, Ruth Bonita, John Raeburn, Martin McKee and Stig Wall (2014) From public to planetary health: a manifesto. *The Lancet*, 383(9920): 847. https://doi.org/10.1016/S0140-6736(14)60409-8

18   Elena C. Hemler and Frank B. Hu (2019) Plant-based diets for personal, population, and planetary health. *Advances in Nutrition*, 10(Supplement_4): S275–S283. https://doi.org/10.1093/advances/nmy117

Chapter 5

1   Carol J. Adams (2015) *The Sexual Politics of Meat: A Feminist-Vegetarian Critical Theory.* 2nd Edition. Bloomsbury.

2   Paloma Cortes (2022) The Bay Area's first vegan Puerto Rican restaurant doesn't skimp on flavor. *KQED*, 19 October. https://www.kqed.org/arts/13920581/vegan-puerto-rican-restaurant-casa-borinquena-oakland-opening

3   Charles Bethea (2023) Special Sauce: The Slutty Vegan burger chain puts the party in plant-based eating. *The New Yorker.* 10 April. https://www.newyorker.com/magazine/2023/04/17/how-slutty-vegan-puts-the-party-in-plant-based-food

4   Barbara McDonald (2000) 'Once you know something, you can't not know it': an empirical look at becoming vegan. *Society & Animals*, 8(1): 1–23, at 1. https://doi.org/10.1163/156853000X00011

5   Sara Ahmed (2014) *The Cultural Politics of Emotion.* 2nd Edition. Edinburgh University Press.

6   Corey L. Wrenn (2023) Shocked or satiated? Managing moral shocks beyond the recruitment stage. *Emotions: History, Culture, Society.* https://doi.org/10.1163/2208522x-bja10045

7   https://www.youtube.com/watch?v=1qiuZE7ysvc&ab_channel=TheIndependent

8   Sparsha Saha (2023) Why don't politicians talk about meat? The political psychology of human-animal relations in elections. *Frontiers in Psychology*. https://www.ncbi.nlm.nih.gov/pmc/articles/PMC10327565/

9   See, for example, Sigal Samuel (2021) It's not just Big Oil. Big Meat also spends millions to crush good climate policy. *Vox*, 13 April. https://www.vox.com/future-perfect/22379909/big-meat-companies-spend-millions-lobbying-climate

Chapter 6

1   Caitlin Welsh, Joseph Majkut and Zane Swanson (2023) The future appetite for alternative proteins. *Center for Strategic and International Studies*. https://features.csis.org/the-future-appetite-for-alternative-proteins/

2   Zafer Bashi, Ryan McCullough, Liane Ong and Miguel Ramirez (2019) Alternative proteins: The race for market share is on. *McKinsey and Company*. https://www.mckinsey.com/industries/agriculture/our-insights/alternative-proteins-the-race-for-market-share-is-on

3   Jan Dutkiewicz and Gabriel Rosenberg (2021) Man v food: is lab-grown meat really going to solve our nasty agriculture problem? *The Guardian*, 29 July. https://www.theguardian.com/news/2021/jul/29/lab-grown-meat-factory-farms-industrial-agriculture-animals

4   Tom Levitt (2021) What's the beef with cows and the climate crisis. *The Guardian*, 27 October. https://www.theguardian.com/environment/2021/oct/27/whats-the-beef-with-cows-and-the-climate-crisis

5   As discussed in this paper co-authored by genomic and social scientists: Neil Stephens, Lucy Di Silvio, Illtud Dunsford, Marianne Ellis, Abigail Glencross and Alexandra Sexton (2018) Bringing cultured meat to market: technical, socio-political, and regulatory challenges in cellular agriculture. *Trends in Food Science & Technology*, 78: 155–166. https://doi.org/10.1016/j.tifs.2018.04.010

6   Garrett M. Broad (2020) Making meat, better: the metaphors of plant-based and cell-based meat innovation. *Environmental Communication*, 14(7): 919–932, at 926. https://doi.org/10.1080/17524032.2020.1725085

7   *VegConomist* (2018) Tyson Foods declare themselves 'sustainable'. 24 September. https://vegconomist.com/company-news/tyson-foods-declare-themselves-sustainable/

8   Pelle Sinke (2021) TEA of cultivated meat. Future projections for different scenarios. Report for CE Delft. https://cedelft.eu/publications/tea-of-cultivated-meat/

9   Caroline Bushnell, Liz Specht and Jessica Almy (2022) Cultivated meat and seafood. *Good Food Institute 2022 State of the Industry Report*. https://gfi.org/resource/cultivated-meat-eggs-and-dairy-state-of-the-industry-report/

10 Dwayne Holmes, David Humbird, Jan Dutkiewicz, Yadira Tejeda-Saldana, Breanna Duffy and Isha Datar. (2022) Cultured meat needs a race to mission not a race to market. *Nature Food*, 3(10): 785–787.

11 The full list of associations is available in the cited article. While there were also some positive associations (clean, healthy), they were overwhelmingly negative: Christopher Bryant and Courtney Dillard (2019) The impact of framing on acceptance of cultured meat. *Frontiers in Nutrition*, 6: 103. https://doi.org/10.3389/fnut.2019.00103

12 Wim Verbeke, Pierre Sans and Ellen J. Van Loo (2015) Challenges and prospects for consumer acceptance of cultured meat. *Journal of Integrative Agriculture*, 14(2): 285–294. https://doi.org/10.1016/S2095-3119(14)60884-4

13 Liz Alden Wily (2011) *Policy Brief: Accelerate legal recognition of commons as group-owned private property to limit involuntary land loss by the poor*. International Land Coalition. https://www.shareweb.ch/site/Agriculture-and-Food-Security/focusareas/Documents/land_policy_%20brief_alden.pdf

14 World Bank (2023) Land. https://www.worldbank.org/en/topic/land

15 Simone Pott (2020) The gap with land ownership is widening around the world. *WeltHungerLife*. https://www.welthungerhilfe.org/news/press-releases/2020/gap-with-land-ownership-is-widening-around-the-world

16 Simone Pott (2020) The gap with land ownership is widening around the world. *WeltHungerLife*. https://www.welthungerhilfe.org/news/press-releases/2020/gap-with-land-ownership-is-widening-around-the-world

17 UK government's 'Land use Statistics, England 2022' which can be found at: https://www.gov.uk/government/statistics/land-use-

in-england-2022/land-use-statistics-england-2022#:~:text=As%20
at%20April%202022%3A,Residential%20gardens'%20(4.9%25)

18   United Nations Food and Agricultural Organization (2020)
Land use in agriculture by the numbers. https://www.fao.org/
sustainability/news/detail/en/c/1274219/

19   George Monbiot (2022) *Regenesis: Feeding the World without
Devouring the Planet*. Penguin.

20   S. Wagura Ndiritu (2020) Beef value chain analysis and climate
change adaptation and investment options in the semi-arid lands
of northern Kenya. *Journal of Arid Environments*, 181: 104216.
https://doi.org/10.1016/j.jaridenv.2020.104216

21   As reported in Olivia Rudgard (2022) Prince Charles backs
face masks for cows to help save the planet one burp at a
time. *The Telegraph*, 27 April. https://www.telegraph.co.uk/
environment/2022/04/27/face-masks-cows-could-help-save-planet-
one-burp-time/

22   Jan Dutkiewicz and Elan Abrell (2021) Sanctuary to table dining:
cellular agriculture and the ethics of cell donor animals. *Politics
and Animals*, 7(1): 1–15.

Chapter 7

1   Jessica Greenebaum (2012) Veganism, identity and the quest for
authenticity. *Food, Culture & Society*, 15(1): 129–144. https://doi.
org/10.2752/175174412X13190510222101

2   Catherine Oliver (2023) Vegan world-making in meat-centric
society: the embodied geographies of veganism. *Social & Cultural
Geography*, 24(5): 831–850. https://doi.org/10.1080/14649365.20
21.1975164

3   Brigid Brophy (1965) The Rights of Animals. *The Sunday Times*.
10 October.

4   Karen M. Morin (2018) *Carceral Space, Prisoners and Animals*.
Routledge.

5   Carol J. Adams (2015) *The Sexual Politics of Meat: A Feminist-
Vegetarian Critical Theory*. 2nd edition. Bloomsbury.

6   Carol J. Adams (2010) Why feminist-vegan now? *Feminism &
Psychology*, 20(3): 302–317. https://doi.org/10.1177/
0959353510368038

7   Aph Ko and Syl Ko (2017) *Aphro-ism: Essays on Pop Culture,
Feminism, and Black Veganism from Two Sisters*. Lantern
Publishing and Media.

8   Nassim Nobari (2023) Meet the new vegan world. In *Oppressive Liberation*. Palgrave Macmillan. https://doi.org/10.1007/978-3-031-15363-1_12

9   Angela Stuesse (2016) *Scratching out a living: Latinos, race, and work in the Deep South*. University of California Press.

10   Jonathan Dickstein, Jan Dutkiewicz, Jishnu Guha-Majumdar and Drew Robert Winter (2022) Veganism as left praxis. *Capitalism Nature Socialism*, 33(3): 56–75. https://doi.org/10.1080/10455752.2020.1837895

11   Andrew Gough (2023) The disturbing link between slaughterhouse workers and PTSD. *Surge Activism*, 24 January. https://www.surgeactivism.org/articles/slaughterhouse-workers-and-ptsd

12   Rasmus R. Simonsen (2012) A queer vegan manifesto. *Journal for Critical Animal Studies*, 10(3): 51–81. https://www.wellbeing intlstudiesrepository.org/cgi/viewcontent.cgi?article=1007&context=acwp_aafhh

13   Rasmus R. Simonsen (2012) A queer vegan manifesto. *Journal for Critical Animal Studies*, 10(3): 51–81. https://www.wellbeing intlstudiesrepository.org/cgi/viewcontent.cgi?article=1007&context=acwp_aafhh

14   Carrie L. Hamilton (2021) Mourning leather: queer histories, vegan futures. *Memory Studies*, 14(2): 303–315. https://doi.org/10.1177/1750698019876001

15   Carrie L. Hamilton (2021) Mourning leather: queer histories, vegan futures. *Memory Studies*, 14(2): 303–315, at 308. https://doi.org/10.1177/1750698019876001

16   Emelia Quinn and Benjamin Westwood (eds) (2018) *Thinking Veganism in Literature and Culture: Towards a Vegan Theory*. Springer.

17   Robert McKay (2018) A vegan form of life, *Thinking Veganism in Literature and Culture: Towards a Vegan Theory*, pp 249–271.

18   Peter Singer (1990) *Animal Liberation. Towards an End to Man's Inhumanity to Animals*. Granada Publishing Ltd.

19   Sunaura Taylor (2017) *Beasts of Burden: Animal and Disability Liberation*. The New Press.

20   Michele Kaplan (2016) Is veganism ableist? A disabled vegan perspective. *Vegan Feminist Network*, 7 September. https://veganfeministnetwork.com/is-veganism-ableist-a-disabled-vegan-perspective/

21   pattrice jones (2013) Intersectionality and animals. *VINE Sanctuary News*, 11 October. https://blog.bravebirds.org/archives/1553

22   Richard J. White (2015) Following in the footsteps of Élisée Reclus. *Anarchism and Animal Liberation: Essays on Complementary Elements of Total Liberation*, pp 212–221.

23   Hal Herzog (2011) Why do most vegetarians go back to eating meat? *Psychology Today*. 20 June. https://www.psychologytoday.com/gb/blog/animals-and-us/201106/why-do-most-vegetarians-go-back-eating-meat

24   Lori Gruen (2014) Facing Death and Practicing Grief, in Carol J. Adams and Lori Gruen (eds) *Ecofeminism: Feminist Intersections with Other Animals and The Earth*. 2nd Edition. Bloomsbury.

Chapter 8

1   Simon Amstell (2017) *Carnage*. BBC Productions.

2   https://www.youtube.com/watch?v=Ny9qvGPx5ds&ab_channel=WePlanet

FURTHER READING AND OTHER RESOURCES

Books

Carol J. Adams, *The Sexual Politics of Meat – 25th Anniversary Edition: A Feminist-Vegetarian Critical Theory* (Bloomsbury Academic, 2015)

Karen Davis, *Prisoned Chickens, Poisoned Eggs: An Inside Look at the Modern Poultry Industry* (Book Publishing Company, 2009)

Eva Giraud, *Veganism: Politics, Practice, and Theory* (Bloomsbury Academic, 2021)

Catherine Oliver, *Veganism, Archives and Animals* (Routledge, 2021)

Sunaura Taylor, *Beasts of Burden: Animal and Disability Liberation* (New Press, 2017)

Laura Wright, *The Routledge Handbook of Vegan Studies* (Routledge, 2023)

Online resources

Open University Open Learn Course on veganism by Matthew Cole: https://www.open.edu/openlearn/veganism

The Vegan-Feminist Network run by Corey Lee Wrenn: https://veganfeministnetwork.com/

Podcasts

The Animal Turn with Claudia Hirtenfelder:
   https://www.theanimalturnpodcast.com/
Brown Vegan with Monique Koch:
   https://www.brownvegan.com/podcast
In a Nutshell from the Plant Based Health Professionals:
   https://plantbasedhealthprofessionals.com/
   the-pbhp-podcast-in-a-nutshell
Knowing Animals with Josh Milburn:
   https://knowinganimals.libsyn.com/
Our Hen House with Jasmin Singer and Mariann Sullivan:
   https://www.ourhenhouse.org/

Cookbooks and websites (tried and tested by the author)

Tim Anderson, *Vegan Japan-easy Classic & Modern Vegan Japanese Recipes to Cook at Home* (Hardie Grant Books, 2020)

Henry Firth and Ian Theasby, *BOSH!: Simple Recipes, Unbelievable Results, All Plants* (HQ, 2018) – especially good for newbies

Marie Kacouchia, *Vegan Africa: Plant-Based Recipes from Ethiopia to Senegal* (The Experiment, revised edition, 2023)

Richard Makin, *Anything You Can Cook, I Can Cook Vegan* (Bloomsbury, 2023)

Isa Chandra Moskowitz, *I Can Cook Vegan* (Abrams, 2019)

Meera Sodha, *EAST: 120 Easy and Delicious Asian-inspired Vegetarian and Vegan Recipes* (Fig Tree, 2019) – entirely vegetarian, mostly vegan or vegan-adaptable

It Doesn't Taste Like Chicken:
   https://itdoesnttastelikechicken.com/
Lazy Cat Kitchen: https://www.lazycatkitchen.com/

Campaigns and charities

The Vegan Society: https://www.vegansociety.com/
Veganuary: https://veganuary.com/
Veggies Nottingham: vegan catering co-operative,
   https://www.veggies.org.uk/

Art, news and media

Just Wondering: short, animated essays,
   https://www.justwondering.io/
Plant Based News: vegan news media and education
   platform, https://plantbasednews.org/
Sentient Media: non-profit news organisation
   https://sentientmedia.org/
We Animals Media: animal photojournalism,
   https://weanimalsmedia.org/

Sanctuaries

Hillside Animal Sanctuary (UK): https://hillside.org.uk/
Sadhana Forest (India, Haiti and Kenya):
   https://sadhanaforest.org/about-us/
VINE Sanctuary (USA): https://vinesanctuary.org/

INDEX

References to figures are in *italics*
and to boxes are in **bold**.